# The Microcontroller Application Cookbook

*Featuring the BASIC Stamp II™*

**Matt Gilliland**

*Forward by* Ken Gracey, *Parallax, Inc.*

**ISBN: 0-615-11552-7**

Note: Information contained in this book has been developed or obtained from sources believed to be reliable. However, the author cannot guarantee the accuracy or completeness of any information published herein and shall not be responsible for any errors, omissions, or damages arising out of the use or misuse of this information. This book is published with the understanding that neither the author nor the publisher is supplying engineering or other professional services. If such services are required, the assistance of an appropriate professional should be sought.



The author can be reached at: matt@miconcookbook.com

This book is available at special quantity discounts to use as premiums and sales promotions, or for use in corporate or scholastic training programs. For more information, please contact the author at: mattgill@jps.net, or call Woodglen Press.

For updates to this book, go to: www.miconcookbook.com

1-888-WOODGLEN

To Kimberly, Madison, Hayley
&
Cooper

## Acknowledgements

Of course I'd like to say that "I did this all by myself", but without the support of the following people, this book would have never been possible.

First, I would like to thank **Parallax, Inc.**, for the invention of the Stamp. This device has radically changed the microcontroller market allowing us "common folk" the ability to create nifty micro-controlled projects using a simple BASIC language. In particular, I'd like to thank **Ken Gracey** for his encouragement and patience during the creation of this book. It was Ken who asked me to write "What's a Microcontroller?" and as an outgrowth of that work, I decided to undertake this project. Thanks Ken!

I'd also like to thank **Mouser Electronics** for providing the components necessary to create the circuits contained herein.

Thanks to **Allegro Microsystems** for their input on Hall-effect devices.

Thanks to **Jane Dickson** and **Linda Gilliland** for turning my poor grammar into something that is understandable!

Thanks also to **Jen Jacobs** for a great cover design.

Most of all, I'd like to thank my wife **Kimberly**. Without her encouragement and "cattle-prodding", this book would never have been.

# Table of Contents

# Forward

# The Main Ingredients...

The Microcontroller Application Cookbook is a collection of interface circuits. For somebody getting started with microcontrollers (referred to as a "micon" by Matt), designing the circuit can be the most difficult part of building a project. Sifting through a myriad of datasheets trying to figure out how to interface a micon can be a challenge to hobbyists, students, and even engineers.

It's quite easy to build a circuit casserole with the collection of ingredients you'll find in this book. Suppose you wanted to automate your greenhouse. Watering solenoid valves could be actuated using a Solid State Relay circuit and a ventilation system might be manipulated with a DC motor controller - both are found in Chapter 4. Circuits with linear temperature sensors, humidity sensors and photocells can be found in Chapter 3. Cook up an application that waters the plants in the morning, opens the greenhouse vents when it's too hot or humid and charges the micon power supply using a solar cell circuit!

In fact, the cookbook approach makes it easier for entrepreneurs to create products. Simply look through the book and pick the circuits that get your project started. Students and educators will find the book to be an instant reference of circuits. For electronic designers it provides a variety of simple circuits and interface code that can be customized for more advanced uses.

### BASIC Stamp as a Kitchen Tool...

This book is not about the BASIC Stamp - it's a collection of useful microcontroller recipes that shows PBASIC code snippets to demonstrate I/O control. The BASIC Stamp's simple instruction set makes it easy to write source code and interface to the circuits. It's the conduit to achieve the circuits. Using the BASIC Stamp is advised for beginners and for those with rapid prototyping needs.

For those of you designing an embedded system, this book provides ready-to-consume circuits for the chip of their choice. For example, parts from Motorola and Microchip have logic threshold transitions and source/sink current capabilities similar to the BASIC Stamp. You would substitute the PBASIC code in this book to the native language used by these manufacturers.

If this book were about the BASIC Stamp, we'd identify it as the most complete single circuit reference available. Our Parallax web sites have an immense amount of application material for the BASIC Stamp, but it details the writing of PBASIC source code more than the hardware circuitry. The Microcontroller Application Cookbook provides a highly accessible reference source for the support most needed by our customers – circuit design.

Matt has also kept it simple with easy-to-find components. Each part in the book is easily obtained from common electronic suppliers. Many of the circuits in this book could be more complex and the accompanying code could be more efficient – but why? At Parallax we've stood in front of classes of teachers trying to teach them how to make quiche when all they wanted to do was bake a cookie. Blinking an LED and generating sound on a speaker is very rewarding when you understand the circuit! Further, most people who use our micon only need the basics to get started with their project.

## Tips from the Line Cooks...

Parallax technical support staff answers a variety of e-mail and telephone calls. Our customers automate their homes, collect weather data on the North Pole, and send micons into space for scientific experiments. Our tech support's review of the Microcontroller Application Cookbook identified several tips we'd like to share with all readers of the book.

Educational customers often want to know how they can protect their micons when used in a classroom environment. Chapter two's "Buffering and I/O Protection" shows you how to size a resistor to prevent damage to I/O pins. Using this advice prevents "letting the smoke out" and could save you $50! Destroying BASIC Stamps gets expensive, not to the mention the inconvenience of running out of them during a project. It's great for our business, but disappointing when students can't learn because the hardware has been destroyed.

Motor control is made easy with Chapter 4's "DC Motor" h-bridge circuitry. Controlling DC motor speed and direction is a frequent topic on our discussion groups. For situations where there's a need to control a high voltage motor for factory automation or robotic projects this section boils an h-bridge down to the basics. I haven't seen a friendly h-bridge circuit until this book.

Avoid getting burned with the high-voltage circuits. The schematics and references provided on power supply isolation and solid state relays demonstrate safe control of high-voltage devices with only a five-volt micon. Work carefully with high voltage and seek advice if needed. Learn to use a voltmeter.

## Bon Appetite!

If you decide to prove some of the circuits with a BASIC Stamp, and still need some help getting started, contact us at Parallax by telephone (916-624-8333), e-mail (info@parallaxinc.com) or stop by our facility in Rocklin, California. We can help identify cost effective ways to get into microcontroller interfacing and find sources for parts. As Matt indicated, there are many helpful people on our newsgroups, so consider joining them at www.parallaxinc.com.

Parallax endorses the Microcontroller Application Cookbook as a collection of well-designed and useful circuits, and we're certain you'll quickly be able to make a little micon do big things after cooking up your own circuits from this useful resource!

Sincerely,

Ken Gracey
Parallax, Inc.

Chapter 1

# Introduction

# A Brief History
*and*
# How to Use This Book

The fact that you've picked up this book indicates you probably know what a computer is, and what it does. In fact, most everyone knows what a computer looks like - it's sitting on your desk with a keyboard, mouse and monitor.

However, if you've been using computers for many years you may remember that they didn't always look this way.

The first computers were "open to the air." The main processor board just sat there on your desk. There was no box to protect the individual components. Many systems didn't have a monitor, and a mouse was something you caught in a trap, not a pointing device.

In those days, programming the computer to "do something" was difficult. Physically connecting the computer to the "outside world" was relatively easy, simply because most computer people were hardware oriented. It was much more difficult to write programs because the art of programming was new, and there weren't many software "tools" to choose from.

As time went on, the computer evolved from an open frame circuit board into a "black box." It became increasingly difficult to connect the computer to the "real world" because it morphed into a data processing device, rather than a hardware controller.

Even a true input/output connection such as a printer, has become a "no-brainer." The hardware connection is a simple cable and, since the software is installed "automatically," the user is given the impression that the printer is just an *extension* of the computer itself.

This ease of connectivity has been wonderful for the office-computing environment. No longer does a secretary or accountant need to worry about flipping DIP switches or resetting shorting blocks back to "factory default."

New developments in software have focused primarily on data manipulation inside the "box," and the "art" of connecting an I/O port to a 60-watt light bulb has faded into obscurity to be practiced only by industrial control engineers or "hardware geeks."

We are now at a point where the computing power on your desktop is capable of charting a starship's course through the Milky Way, but most "computer" people cannot connect a simple LED to that same computer and make it blink on and off.

Of course, most people who receive some sort of computer training end up with a job that operates on data contained *inside* the "box." Unfortunately, these people may never experience the great sense of accomplishment of making a computer *do something physical in the "real world."*

Several years ago, Parallax developed a microcontroller called the BASIC Stamp. Not only is it a great learning device, but many commercial products have been designed around it as well. The Stamp has helped bridge the gap between hardware connectivity and programming ease.

The Basic Stamp I and Basic Stamp II microcontrollers are quite simple to connect to the "real world" and more important, *very easily programmed* with a specialized version of the popular BASIC programming language (called PBASIC). Using a microcontroller like the Stamp, it is now possible to develop fun and fascinating "machine controller" projects without any formal hardware training.

So why this book?

Hobbyists and professional engineers alike have created thousands of different types of projects with microcontrollers (such as the Stamp).

However, it's interesting to note that most "computer people" seem to have far less difficulty in writing a control program than building the hardware portion of the project.

In other words, many find it easy to *write a program* that will blink a lamp on and off, but they're *not sure what parts to use or how to assemble* the "blinker" circuitry.

This book is written for those of you who know what you want to accomplish and have a pretty good idea of how you're going to write the control program, but need a resource on how to *physically* put the circuit together.

By definition then, this book is *not* a collection of experiments or projects. Rather, it is intended to be a resource for hardware interfacing techniques and circuits geared toward microcontroller applications.

Microcontrollers are amazing devices. The same chip can control the automatic door at the supermarket, receive and decode data on your cellular phone, or analyze soil samples on a distant planet. The silicon remains the same; the only differences are the code (*your* program) and the sensors or output devices connected to it.

Applications for microcontrollers are limitless. In fact, at any one time thousands of new products are under development that take advantage of both its low cost and flexibility.

The Basic Stamp II was chosen as the basis for this book for two reasons.

First, it is *extremely* easy to use and program. Programs can be written, debugged, and tested in a matter of minutes with virtually no development system (other than a personal computer and some *free* editor/compiler software). Therefore, the Stamp is an effective way to provide you with sample code (demonstrating how the circuits work), without getting bogged down in sometimes difficult to understand languages.

Secondly, the Stamp has the same electrical characteristics as most other types of microcontrollers. Therefore, hardware circuits developed for the Stamp can be easily implemented on other microcontrollers with little or no modification.

For simplicity, throughout this book, whenever I refer to the "Stamp," I'm referring to the BASIC Stamp II. Although the BS-1 is still available (and probably will be for quite some time), the enhanced command set and additional memory capacity of the BS-2 make it my preferred choice.

The Stamp is a very elaborate blend of hardware and software. Taken *individually*, the components that make up the Stamp are complex, but *together* they form a sophisticated, yet easy to use electronic brain.

I'm also going to take the liberty of introducing a new term – "*micon"* – a shortened form of the word MIcroCONtroller. I use it in "the lab," after getting tired of saying "microcontroller" all day long.

In this book you'll find many different circuits that have been specifically adapted for microcontroller applications. Although the program samples are written for the BASIC Stamp II (in "PBASIC"), the hardware interfacing techniques apply to most other types of microcontrollers. You'll simply have to change the code.

I'm assuming a couple of things about you, the reader. The first is that you have some knowledge of basic electronics. For example, you know what a resistor is, and what it does. Secondly, I assume that you've had some experience at assembling electronic circuits, i.e. you know how to solder components onto a breadboard and can read a schematic diagram.

If you don't have these skills, there are some excellent resources available to help you "get up to speed". Among these are:

*Programming & Customizing the BASIC Stamp Microcomputer*
Scott Edwards, TAB Books ISBN #0-07-913684-2.

> Not only are there some fascinating (Stamp based) projects in this book, but the first half is an excellent beginning text on how to assemble and work with electronic circuits.

The Parallax website at: www.parallaxinc.com

More resources than I can mention here. This is the place to go if you're a Stampaholic. They also maintain a "discussion list" that has a solid following of Stamp experts and novices alike.

The "Stamps in Class" website: www.stampsinclass.com

Another official Parallax website, but with more of an educational focus. Some great learning materials here, available for free download.

Microcontrollers are designed to make decisions based upon sensory inputs. Therefore, Chapter 3 includes many types of sensors and input devices that you might want to connect to a microcontroller.

After the micon makes its decision, output capability is required for "real world" manipulation. Chapter 4 covers common interface techniques for many types of output devices.

As far as software is concerned, you're *not* going to find a complete program in this entire book. With each circuit schematic you'll get just enough code to deliver the data to or from the output devices or sensors.

I believe (because you've read this far) that you have already decided what type of device (or product) you want to create, and my intention is to provide those hardware and software "snippets" of information that can become integral "building blocks" for your project.

One of the main reasons that the Stamp has had such great success in the microcontroller market is because it's so easy to program. In order to facilitate quick adoption of the device, Parallax developed its own version of the BASIC programming language.

Called PBASIC, it concentrates on I/O operations rather than on other aspects of what I call "screen programming." Screen programming is when the primary intent of the computer program is to manipulate and display data (using a monitor) or information. Examples of this might be a video game, word processor, or a spreadsheet program.

Because PBASIC concentrates on input/output operations, the instruction set may seem more restrictive than what you might be accustomed to. However, remember that microcontrollers are not really designed to interface with humans, but instead are optimized as "machine controllers."

If you don't have it already, go to the Parallax website and get yourself a free copy of the *BASIC Stamp Manual.* It will be a helpful reference as we go through the pages that follow.

Remember, the complete circuit design (for your project) and the required programming skills are your responsibility. In the following pages you'll only get a schematic diagram and a code "snippet" – just enough to demonstrate how the circuit works.

So whether it's a "solar powered automatic remote weather data logger" or your own version of "R2D2," you'll find many useful circuits herein, but the actual construction and coding is up to you.

This is a cookbook with lots of ingredients, not all of which you'll use on any one project. But sooner or later if you're like me (many people would say that's *unfortunate*) you'll wonder, "What if I connect a... to a... *hmmm - where's that soldering iron?"*

Enjoy!

Chapter 2

# Hardware Considerations

# Microcontroller Fundamentals

Microcontrollers are simply computers that are designed to interface with machines or devices rather than with people. Because of this, they don't need a monitor, keyboard, mouse or any other device that requires human interaction.

Although it can be done, it is significantly more difficult to connect a monitor or keyboard to a microcontroller than it is to attach a similar "plug-n-play" device to your PC.

Of course, in order to create a "control program" we (humans) need these "user interfaces" to communicate with the microcontroller. That is the purpose of a development system.

Microcontrollers themselves are usually quite inexpensive (as low as a couple of dollars), when compared to the price of a development system. The system can cost hundreds or even thousands of dollars depending upon its level of sophistication.

The Stamp II, in this case, goes against the tide. As of this writing (early 2000) it costs about $50.00 - somewhat expensive for a microcontroller. However, for this increased cost you get a simple-to-use device that can be easily programmed in a language called "PBASIC." And as an added bonus, the development system can be free (if you're willing to work a little).

The Stamp also has a huge amount of additional information available. If you're looking for applications help or interfacing and programming techniques, be sure to check out the Appendix for more resources.

A development system consists of some method of connecting the micon to a personal computer (usually via a serial port), and some sophisticated software. If you choose to “build” rather than “buy,” Parallax has provided the wiring diagrams necessary to create your own “Carrier Board” and “Programming Cable” with commonly available components (See Figure 2.2). Check your "junk drawer" first - you may already have many of the necessary parts.

The Stamp “Development System” software is available for *free* from the Parallax website, as is an up-to-date version of the entire *BASIC Stamp Programming Manual.*

Once you have these items on your bench you’ll be writing your first control program in no time. It’s that easy!

The choice of “expensive microcontroller and free development system” *-or-* “cheap micon and expensive (and *perhaps* more difficult to use) development system” is really dependent upon your application. Will it be a production item? Is it a limited production run? How much time do you want to spend on the “learning curve?” These decisions are yours to make.

There are many different microcontrollers from which to choose, and they all have slightly different features and capabilities. Your project will dictate which to use, depending upon circuit requirements, level of sophistication, and cost.

Most micons have a CPU, RAM, ROM and (in the case of the Stamp) some EEPROM (Electrically Erasable Programmable Read Only Memory).

Inexpensive controllers that are designed to be programmed once (for a particular application) may omit the EEPROM, because there may be no need to ever change the program (once the product goes "into production").

The Stamp II holds your program in a two kilobyte EEPROM, which allows you to erase and reprogram it *millions* of times.

The Stamp II is also unique because it contains a full PBASIC interpreter residing in ROM located right on the Stamp chip itself. I say "chip," but if you look closely it's not so much a chip as it is a printed circuit board containing *multiple* chips. See Figure 2.1.

**Figure 2.1**
***The BASIC Stamp II Microcontroller***
*Courtesy of Parallax, Inc.*

The largest integrated circuit on the Stamp is the actual microcontroller - a PIC series micon made by Microchip Technology, Inc. Within this PIC resides a full PBASIC interpreter developed by Parallax, Inc. This code is burned into ROM, and is permanent (non-erasable).

There's also a two-kilobyte EEPROM that holds your program and any data storage required in your application. The EEPROM can be erased and re-programmed about 10 million times, so unless your program collects and writes data to memory constantly, there's no need to worry about "wearing out" the EEPROM.

Other items include a power-up reset circuit, voltage regulator, system clock, and some components for communicating over a serial connection.

See the *BASIC Stamp Manual* and/or *Programming and Customizing the BASIC Stamp Computer* (by Scott Edwards) for more complete descriptions of the Stamp circuitry.

Figure 2.2 is a schematic of the BASIC Stamp II serial port interface to a personal computer's DB-9 serial COM port. This is the diagram you'll need, should you decide to build the hardware portion of the development system.

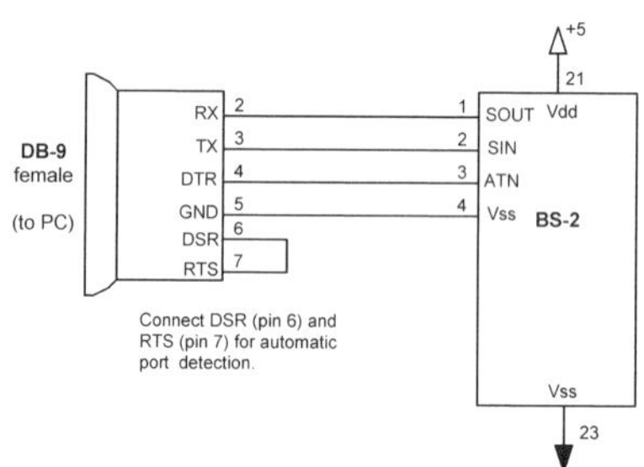


**Figure 2.2**
***Programming cable connections from a PC to the BASIC Stamp II***

There are other tools that you may find useful during the prototyping phase of your project. The Parallax website is a great place to find accessories such as the "Board of Education" (BOE) which make it quite easy to prototype circuits.

The BOE has a small "experimenter prototyping area" attached to a printed circuit board that contains a BS-2 socket, as well as the DB-9 serial port connector (for attaching to your personal computer).

For more complicated projects you may wish to create your circuit on a breadboard, or even etch your own printed circuit board.

In any case, interfacing the Stamp to the PC must be through the circuit connections as shown in Figure 2.2.

If you do decide to create your project on a printed circuit board (PCB) or perf-board, you may not want to include the DB-9 (male) connector (like on the "Stamp Carrier Board," or the BOE) since it will probably only be used during the development phase of your project. The DB-9 connector and circuit board "real estate" would be wasted.

Figure 2.3 shows a programming cable that I've found useful during the construction and development of my projects. Instead of using a DB-9 to connect to the Stamp, it uses a 24 pin DIP Clip (sometimes called a "glomper clip"). These are available from several different manufacturers. The one I use is made by 3M and costs about $12.00.

This cable is useful because you clip it onto the Stamp only while developing the software portion of your project. When the system is ready to "solo," you simply "un-clip" the cable. There is no wasted PCB space, or any unused components.

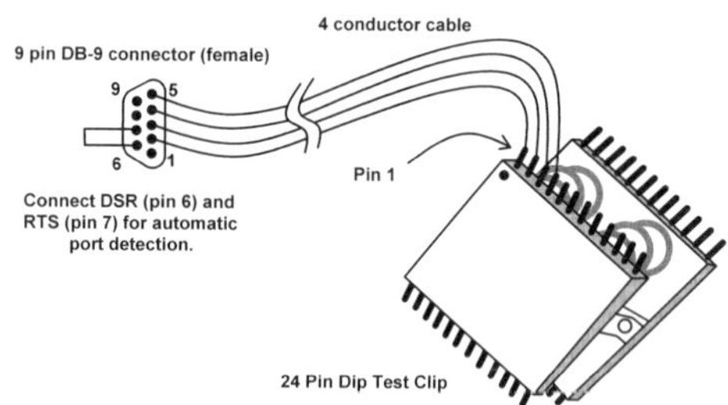


**Figure 2.3**
***The "DIP-clip" Stamp II programming cable***

# Power Requirements

Most microcontrollers operate on a +5 volt dc power supply and use very little current - typically less than 10 milliamps. The Stamp II runs on about 8 milliamps and, when in different "power-down" modes, will typically draw about 50 *microamps.* Of course, when looking at power supply sizing, you must take into account all of the additional circuitry that is connected to the micon, as this is usually where most of the amperage will be required.

For example, let's suppose that you connect a single LED to an I/O line on the Stamp. With a current limiting resistor of 240 ohms, the LED alone will draw more than 10 milliamps - more than the typical current requirement of the Stamp itself. Add a few more components like this to the circuit and the current drain of the microcontroller itself becomes inconsequential.

Although the amperage required by the micon is minimal, you must ensure that the power is "clean" and free of voltage fluctuations. A regulated power supply is a must, but in many cases may not be adequate for a glitch free system.

If your project includes any "heavy current" devices such as motors or solenoids, you will avoid many troubleshooting headaches by isolating the power sources from these elements.

Figure 2.4 shows how you can deliver isolated power to two different portions of a typical circuit.

Notice that the positive supply lines are kept separate from each section of the circuitry. However, in order for the circuit to function properly, the grounds from each power source must be connected.

Inductive loads (such as a motor or solenoid) may generate voltage spikes (called "glitches") during their operation. Using a power supply circuit like this will eliminate most of these fluctuations.

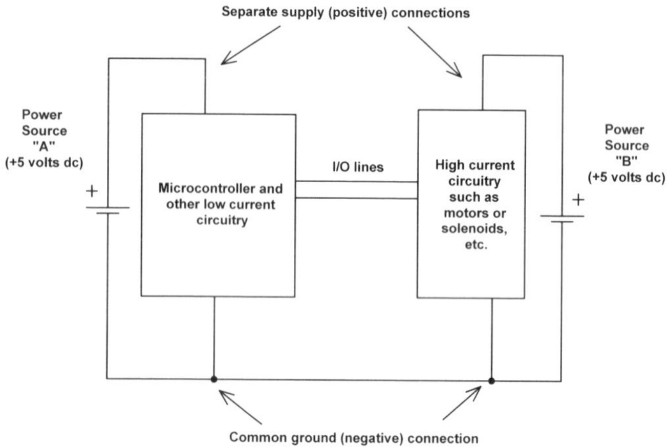


**Figure 2.4**
***Positive power supply isolation in a microcontroller circuit***

In extreme cases however, you may need to *completely* isolate the sensitive computer control circuits from the high current sections. This would require *no* common electrical connections between the two different portions of the circuit.

Figure 2.5 also uses two separate power sources but notice that there is no common connection between the sets of power supply lines.

This circuit uses an optoisolator on each of the I/O control lines resulting in *no common electrical connection* between the independent circuits. The optoisolators use a *beam of light* to control different functions through the I/O portion of the circuit. See Chapters 3 and 4 for some typical circuits that interface to optoisolators.

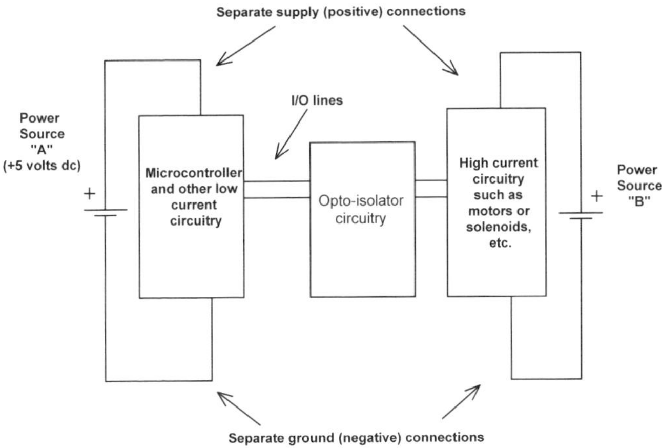


**Figure 2.5**

***Completely isolated power supplies in a microcontroller circuit***

Unlike many other microcontrollers, the Stamp has a built-in 5-volt regulator. You can connect an unregulated power source as shown in Figure 2.6, allowing the regulator to supply power to the Stamp and some limited circuitry. The total current that the on-board regulator can provide is limited to 50 milliamps.

For all but the simplest projects this is somewhat restrictive, so you may wish to use the circuit in Figure 2.7. This circuit utilizes an external voltage regulator (such as the LM7805), which can deliver significantly more current.

The LM7805 is available in several different package styles (TO-92, TO-220, etc.), allowing from 100 ma to several amps of output current. Be sure to check the "pin out" diagrams of the device you select, as the pin locations do vary depending upon the package design.

Don't forget to use adequate filter capacitors on both the input and output lines of the regulator. It's also good design practice to place a bypass capacitor (.1 mf) as near to the positive power pin on each chip in your circuit, especially if you're using any TTL integrated circuits in your project. This practice helps to eliminate noise and transition "glitches" in the system. Clean power is a must!

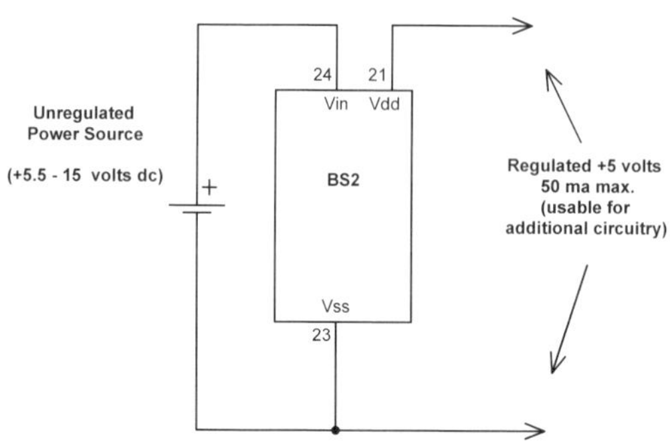


**Figure 2.6**

***Using the built-in +5 volt regulator on the BASIC Stamp II***

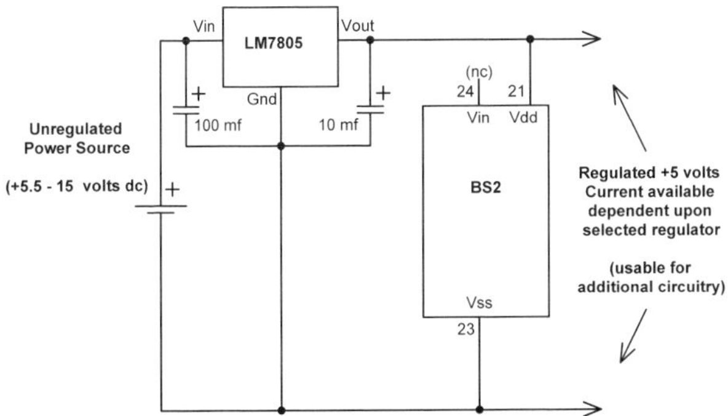


**Figure 2.7**

***Adding an external +5 volt regulator to a BASIC Stamp II***

# Buffering and I/O Protection

Most microcontrollers (including the Stamp) are MOS type devices. Therefore, be careful of static discharge when handling and assembling your circuitry. Although not absolute necessities, a wrist strap and "zero voltage tip" soldering iron may save you money, by preventing damage caused by Electro-Static Discharge (ESD).

Dry windy weather, carpeting, and inappropriate clothing will cause erratic circuit behavior and probable damage to the sensitive circuitry. It's *not* fun when a $50.00 chip "blows". The money is one thing, but the frustration of troubleshooting an inoperative circuit (or program) because of an ESD damaged chip can take its toll on even the most avid inventor!

If you're using the Stamp II for your project, you *are* working with a $50.00 chip. Careful!

It may also make sense (at least during the experimental stage) to place a 270-ohm resistor in series with each of the I/O pins. This will prevent excessive current from flowing through the port, should a "mis-connect" or program glitch occur.

Although the rest of the circuits in this book do not show a series resistor on any of the I/O lines, you may wish to incorporate this protective feature, as shown in Figure 2.8.

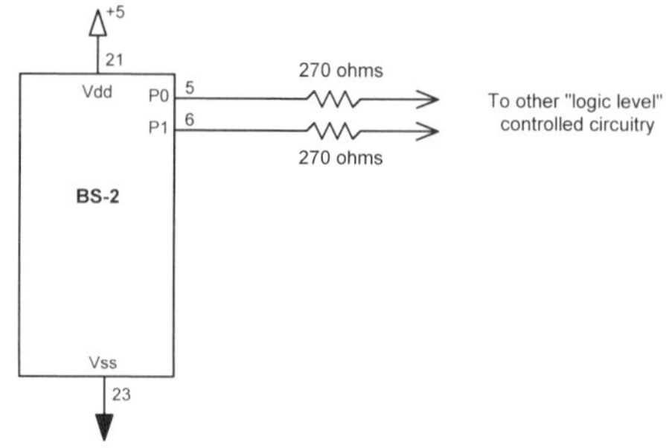


**Figure 2.8**
***Protecting the I/O lines of a Stamp with resistors***

If you're going to dedicate a particular I/O line as an input, it might be a good idea to use the resistor, just in case your program "inadvertently" causes the pin to become an output.

For example, suppose you had the circuit as shown in Figure 2.9. Any time you press the switch, P0 is connected directly to ground. If your program (during the development phase) "accidentally" caused P0 to become an output and go "high", P0 would try to deliver a +5 volt logic level through a direct short to ground, possibly destroying the driver circuitry located inside the Stamp.

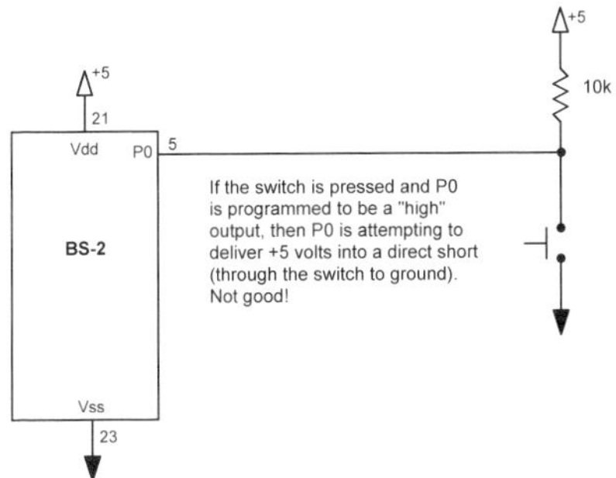


**Figure 2.9**
***Potentially damaging circuit if your program "accidentally" makes the I/O line an output***

However, as shown in Figure 2.10, the resistor limits the amount of current to: 5 volts divided by 270 ohms, or about 18 milliamps - within the specified 25-milliamp maximum of the Stamp.

Notice that the circuit will still operate as desired because the 270-ohm resistor is an insignificant value when compared to the 10K pull-up resistor. When the switch is "open", P0 is pulled high by both resistors (10K and 270 ohms).

When the switch is pushed, the 10K resistor is taken (almost) completely out of the circuit - other than a small current flow from +5 to ground through it. P0 is connected to ground through a minimal 270-ohm resistor – easily "low" enough of a signal to be recognized as logic "0" by P0.

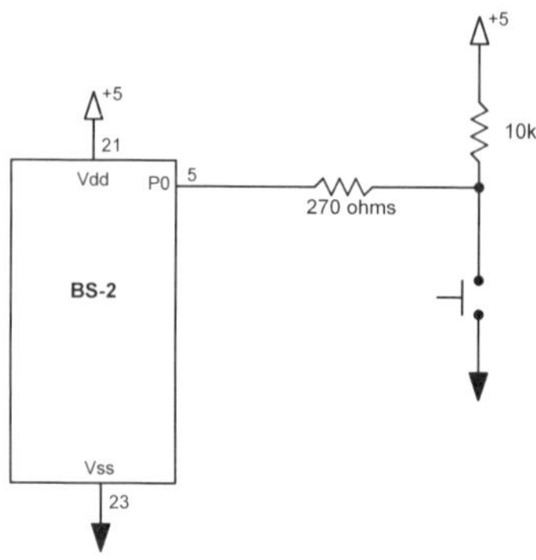


**Figure 2.10**
***The I/O line is protected from programming "accidents"***

The Stamp should not be allowed to source or sink more than 25 milliamps on each I/O pin.

There is a further restriction however, that each 8 bit group (P0-P7 and P8-P15) should not try to source more than 40 milliamps or sink more than 50 milliamps *in total.*

Suppose that you wish to connect 8 LED's to each of the I/O pins (P0-P7). The I/O line going low, or "sinking to ground" turns each of the LED's ON.

Since we're limited to a total of 50 milliamps for the entire group of 8 lines, the current through any one I/O line should not exceed 50 milliamps divided by 8, or 6 milliamps.

This may result in the LED's not being as bright as you would like. But to prevent the I/O port from damage you should not use anything less than 820 ohm resistors, as shown in Figure 2.11.

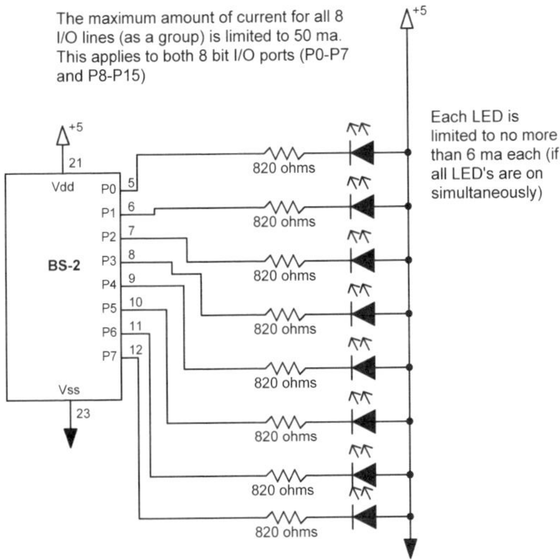


**Figure 2.11**
***Current limits on the BS-2 I/O lines***

To increase the intensity of the LED's, we need to allow more current to flow through them. Since the Stamp is not (safely) capable of delivering any more current, you may need to "buffer" the I/O lines as shown in the next circuit.

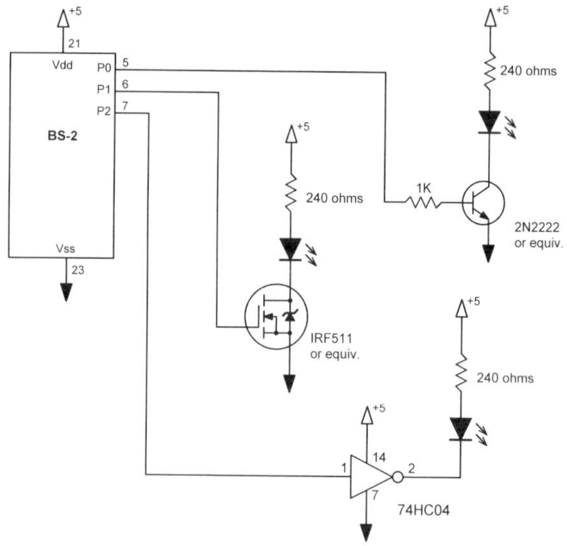


**Figure 2.12**
***Various buffering circuits that provide higher drive capability***

If a particular I/O line will always be used as an output, but 20 milliamps is not enough current for your particular application, you can add a higher current "driver" to the circuit.

There are many different circuits that can be used to increase the drive capabilities of a micon. Three of the more common methods are shown in Figure 2.12.

The circuit that is connected to P0 utilizes an NPN transistor. Bipolar transistors, such as the 2N2222 are current actuated devices – they must have a current flow from the Base to the Emitter in order for the transistor to conduct. Although a bipolar transistor may not be as efficient as the other methods, they are inexpensive and readily available.

Note that in Figure 2.11, the I/O lines must be “low” for the LED to turn on. The buffer circuits shown in Figure 2.12 act as “inverting” buffers. That is, the LED will glow when a “high” signal is presented on the I/O line. This could be desirable, as it may be easier to think of turning something “on” by causing the I/O line to go “high”.

The second buffer circuit (connected to P1) uses a MOSFET type transistor. It costs a little bit more that the NPN but has the ability to supply large amounts of current (sometimes measured in *amps*). Therefore, if you wanted to use an incandescent lamp instead of the LED, simply remove the LED and its current limiting resistor, and replace it with the lamp.

A MOSFET is a voltage-controlled device, therefore there is no significant current flow required to “turn on” the device, causing it to conduct. Also, there is very little voltage drop across the MOSFET, which means almost all of the power used within the circuit is used by the “load” (in this case, the lamp).

The final buffer circuit (connected to P2) uses a commonly available inverting buffer IC. The convenience of this type of device is that a single 14 pin IC contains six independent buffers.

# Level Shifting and Signal Conditioning

Since the Stamp (like many other microcontrollers) is designed to operate on no more than 5 volts, it is important that the I/O lines are not connected to voltages that could exceed this amount.

For example, Figure 2.13 shows a simple switch input circuit. You can see that if the switch is pushed, it will connect the Stamp's I/O line (P0) directly to 12 volts. This exceeds the maximum recommended input specifications for the micon and may result in permanent damage. It is therefore necessary to limit the input voltages (to the micons I/O lines) to no more than the maximum allowed (in most cases, +5 volts dc).

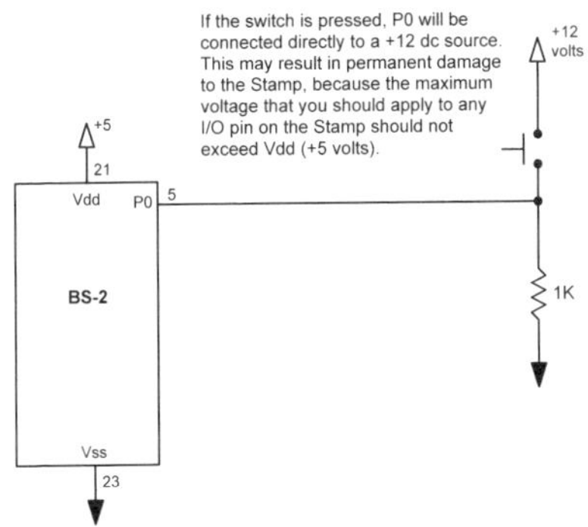


**Figure 2.13**

***Too much voltage on an I/O pin will damage the microcontroller***

Figure 2.14 shows an example of how this high voltage can be "clipped" down to an acceptable level. The circuit uses a zener diode to regulate the voltage down to no more than 5.1 volts dc on the I/O pin.

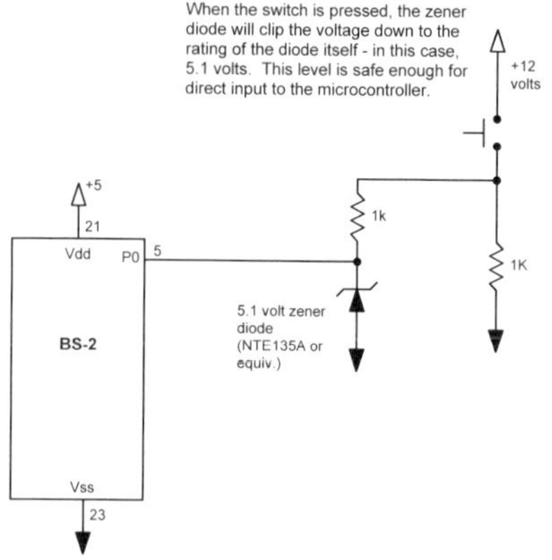


**Figure 2.14**
***Zener diode input voltage protection***

Another method is to use a simple voltage divider, as shown in Figure 2.15. This circuit assumes that the high voltage input remains constant.

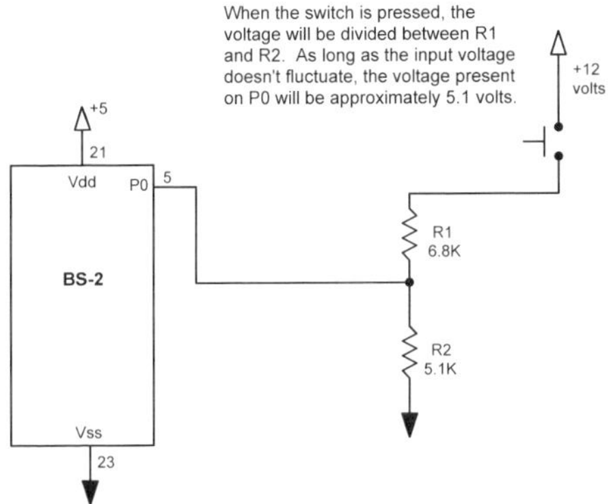


**Figure 2.15**

***A simple resistive divider to decrease the input voltage to a safe level for the micon***

In this example, the 12-volt input is "divided" by the 6.8K and 5.1K resistors, resulting in 5.1 volts presented on the I/O pin. If your input voltage is different, use the following formula to calculate the values of the resistors:

Vout = (Vin) x (R2 / R1+R2)

The simplest method is to select a value for R2. Then, knowing the input voltage and desired "divided down to" voltage (+5 volts), solve the equation for R1.

Suppose that the input voltage is 24 volts DC. The input to the I/O line can be no higher than 5 volts. Arbitrarily select a value of 5.1K for R2.

Then, solving for R1, we have:

5 volts = (24) x (5.1K / R1 + 5.1K)
R1=18.8K

Of course, since an 18.8K resistor is not a common value, we can get pretty close to our target by using a 10K in series with a 9.1K, as shown in Figure 2.16.

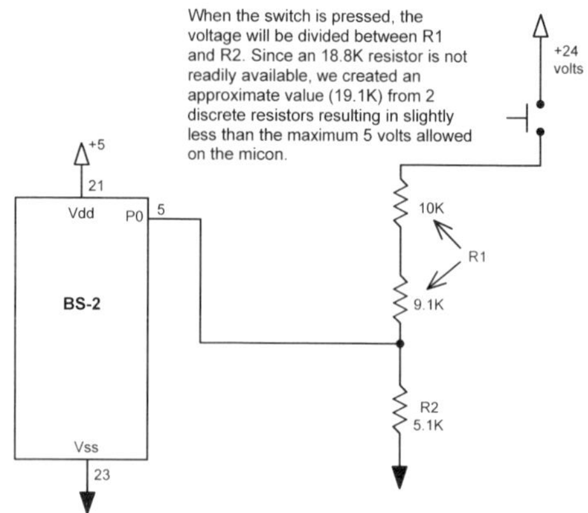


**Figure 2.16**
***Creating the right value resistor for the appropriate voltage input***

If the input voltage is too low for a microcontroller to detect, then you will need to amplify the signal.

Suppose that your input voltage goes from 0 to 1 volt. Whenever anything over .7 volts is present, you want it to be detected as logic “high”.

Since a microcontroller’s typical threshold level for sensing a “high” is about 1.4 volts, the circuit will not operate properly. See Figure 2.17.

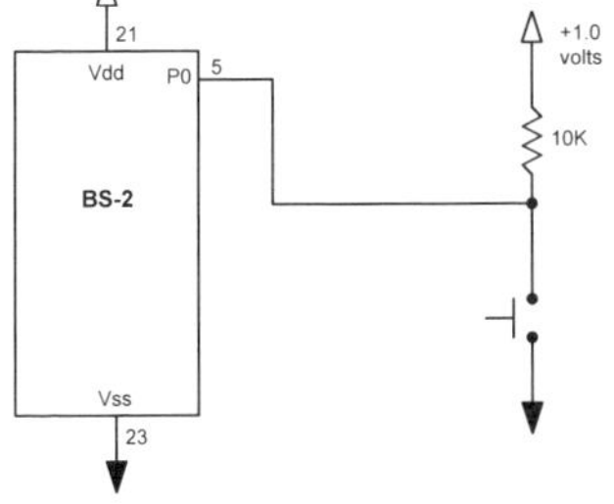


**Figure 2.17**

***Not enough voltage available on the I/O pin will result in NO detection of the signal***

By using a comparator circuit as shown in Figure 2.18, whenever the input voltage exceeds the setpoint (as determined by the voltage on pin 4 of the LM339), the comparator will present a logic "high" to the input of the microcontroller.

With the values shown, the comparator will output a "high" when approximately .7 volts is exceeded on the input. The micon will now detect about 3.4 volts (the output of the LM339), which is easily high enough to be recognized as a logic "1".

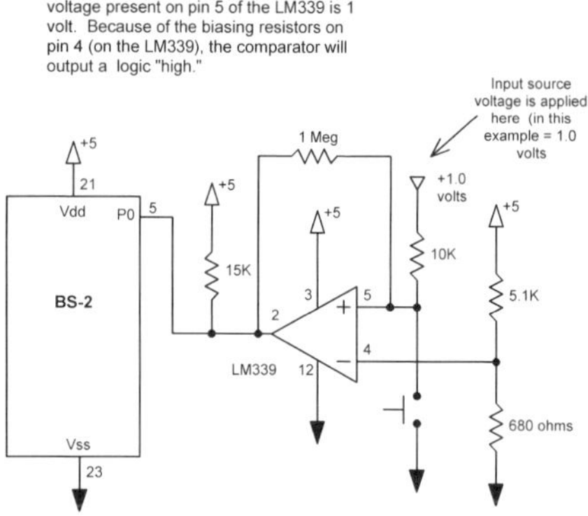


**Figure 2.18**

***Using a comparator to shift the voltage input to a detectable level***

These examples of voltage level shifting (up and down) are only scratching the surface of available techniques. Your circuitry might be a little different, and may require slight modifications to achieve desired results.

# Chapter 3
# Input

# Switches

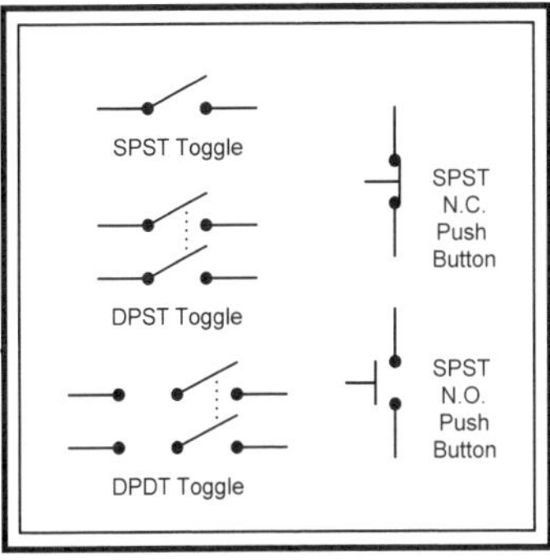


Switches come in a wide variety of shapes and sizes.

However, they can be divided into two major categories: "momentary" & "permanent." Although the basic action (opening or closing a circuit) is the same, the selection of a switch type will depend upon your application.

The following circuits will operate with both types of devices, but your code may need to be modified (depending upon the application). The microcontroller really doesn't know (or care) which type of switch you use, so long as the contacts are clean enough to provide a good signal.

A *momentary* switch does exactly what its name implies – opening or closing of the circuit is made only during its actuation. It goes back to its "normal" state (this could be "normally-open" or "normally-closed" – abbreviated N.O. and N.C. respectively) when the actuation force is removed. These types of switches are typically used as sensors (such as detecting when a door has closed, etc.), or as keypad inputs.

A *permanent* switch is one that requires separate actions to both open and close the circuit. A standard wall-mounted light switch readily demonstrates this. You must flip the switch to turn ON the light, where it will remain until you flip it OFF. The two-position version is also known as a "bi-stable" switch.

Momentary and permanent switches come in many configurations. They can be in the form of a "toggle" (like the wall-mounted light switch), push-button, rotary, rocker, or even a spring-return pressure switch from a well's pressure tank.

In its simplest form, a switch is nothing more than a set of contacts that open or close a circuit. The most basic switch, called an SPST (for "single pole, single throw"), has one set of (two) contacts. When you flip the switch, the contacts either close (completing the circuit) or open (breaking the circuit).

A DPST (double-pole, single-throw), also does what its name implies – with a single flip (or "throw"), two different sets of contacts (called "poles") are actuated, either opening or closing the circuits.

An SPDT (single-pole, double-throw) has three contacts. One of these is "common" to the switch. This common terminal will be connected to either one of the other two contacts – but not to both simultaneously (generally speaking – see below).

For example, suppose the switch has three contacts labeled "A", "B" and "C," and (according to the data sheet) terminal "B" is the "common." With the switch in one position, a circuit is made between A and B. When the switch is moved to the other position, the A-to-B circuit is broken and the B-to-C circuit is made.

DPDT (double-pole, double-throw) is a combination of the above examples: two completely separate sets of (3) contacts. Each set of three contacts operates just like an SPDT type, but the sets work in tandem.

Triple-pole, Quad-pole, etc. are less common types but still operate in the same manner, they just have more contacts.

Another consideration is whether the switch is a "break-before-make" or a "make-before-break" type. In the case of an SPDT type switch, a "make-before-break" would indicate that when you flip the switch from one position to the other (A-B connection to B-C connection), for a brief period of time A, B and C terminals are all connected together. That is to say that the B-C connection is made *before* the A-B connection is broken.

In certain applications this may be a desirable feature, but exercise caution because it is quite easy to create a short circuit condition with this type of switch.

For example, if you connected Vcc to terminal "A" and ground to terminal "C" (to switch either a high or low signal into a microcontroller's I/O port), in that brief period of "flipping," Vcc is connected directly to ground. The connection between B-C is made before the A-B connection is broken, resulting in a brief short circuit between A and C.

Unless specified otherwise, most switches are the "break-before-make" type, and are far more common and readily available.

A simple switch input circuit is shown in Figure 3.1. This is an "active low" circuit. That is, when the switch is pressed, the signal presented to the micon is a "0."

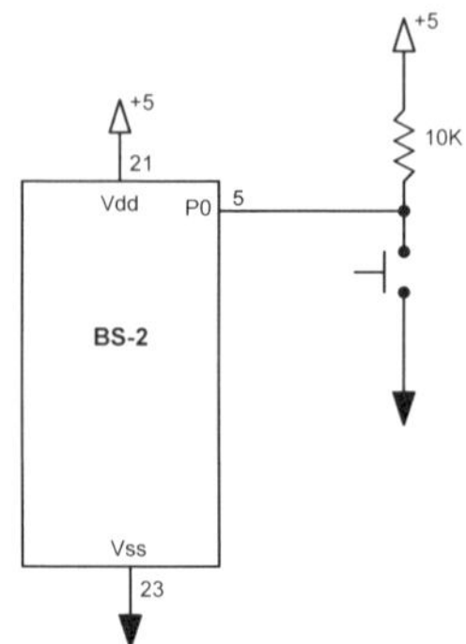


**Figure 3.1**
***An "active low" switch input circuit***

**Code 3.1**

```
here:
if in0=0 then there
goto here
there:
debug "switch pushed"
goto here
```

The "normal" position of the switch is open; therefore P0 is pulled "high" by the 10K resistor to +5 volts. When the switch is pushed, P0 is connected to ground, resulting in a "low." A small current flows through the resistor to ground.

Remember to use caution when in the development phase of your project. It may be a good idea to place a protective current-limiting resistor in series with the I/O pin to prevent a direct short to ground if your program inadvertently causes P0 to become a "high" output. See Chapter 2 for more information.

Fig. 3.2 is an "active high" version of the same circuit.

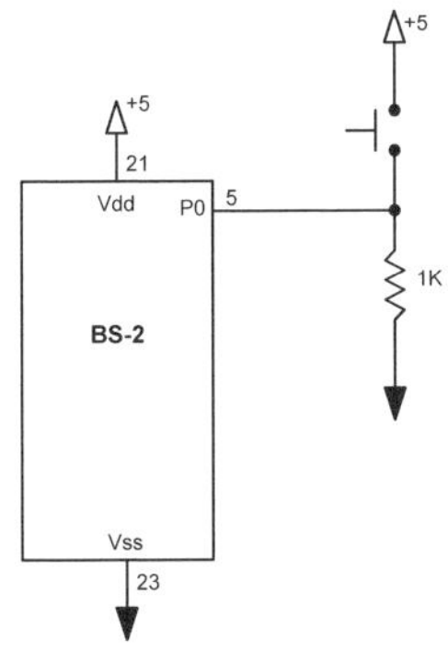


**Figure 3.2**
***An "active high" switch input circuit***

**Code 3.2**

```
here:
if in0=1 then there
goto here
there:
debug "switch pushed"
goto here
```

When the switch is pushed, this circuit will draw a bit more current, due to the lower value of the resistor (1K).

When using switches as input devices, it is usually better design practice to use the circuit in Figure 3.1 than it is to use Figure 3.2. The foremost reason is "noise" prevention.

The threshold voltage to detect a low (logic level "0") is 1.4 volts or lower. This means that when an I/O pin is pulled high to Vdd (+5 volts), it must drop at least 3.6 volts (via a pushed switch) to be recognized as a low.

Conversely, as shown in Figure 3.2, P0 is pulled down to ground through the 1K resistor. In order for the I/O line to recognize a "pushed switch" as a logic "1," the voltage only needs to rise 1.4 volts (the threshold level).

The "pull-up" version of this circuit results in an additional 2.2 volts of "noise threshold". This simply means that Figure 3.1 is less likely to allow voltage spikes (know as "glitches") to disrupt your circuitry.

You can reverse these circuits to operate with normally-closed switch types, as shown in Figures 3.3 and 3.4.

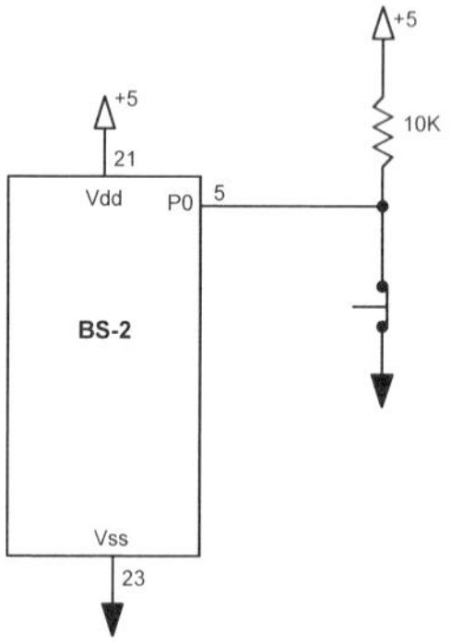


**Figure 3.3**

***A normally-closed, "active high" switch circuit***

**Code 3.3**

```
here:
if in0=1 then there
goto here
there:
debug "switch pushed"
goto here
```

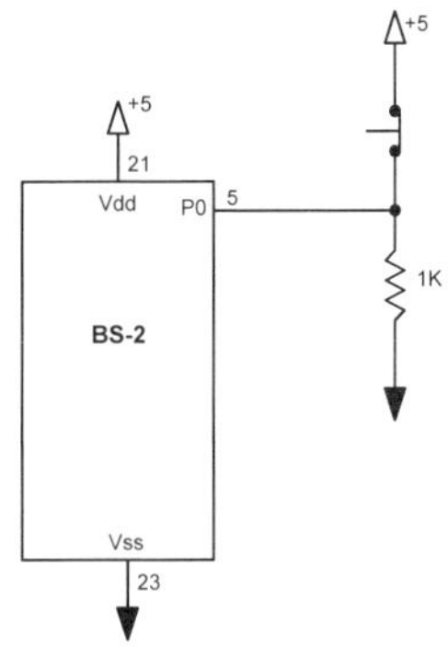


**Figure 3.4**
***A normally-closed, "active low" switch circuit***

**Code 3.4**

```
here:
if in0=0 then there
goto here
there:
debug "switch pushed"
goto here
```

Mechanical switches suffer from a condition known as "switch bounce." This simply means that when the contacts first close, they don't make a clean and instantaneously permanent connection. The switch "chatters" for a moment, resulting in multiple logic-level transitions.

If your microcontroller happens to be looking at the switch during these oscillations, it may not get a true indication of the switch's current state. Therefore, you may need to eliminate these oscillations and provide a "clean" signal to the microcontroller.

The Stamp II has a built-in command called "Button." It is a software routine that "eliminates" switch bounce. Button requires 250 microseconds to execute, so if your timing isn't critical, it's the simplest solution. See the *Basic Stamp Manual* for more information.

Figure 3.5 is an LM555 timer de-bounce circuit. It provides a clean signal whose duration is controlled by R1 and C1.

This circuit is also useful if the switch action happens very quickly, and you need to "stretch" the duration of the signal to allow the micon time to detect it.

For example, suppose your program needs to accomplish certain tasks and can't always be looking for this particular switch input. This circuit allows your program time to attend to other tasks, and still be able to detect when a switch action has occurred. The circuit resets itself after the timing cycle is completed.

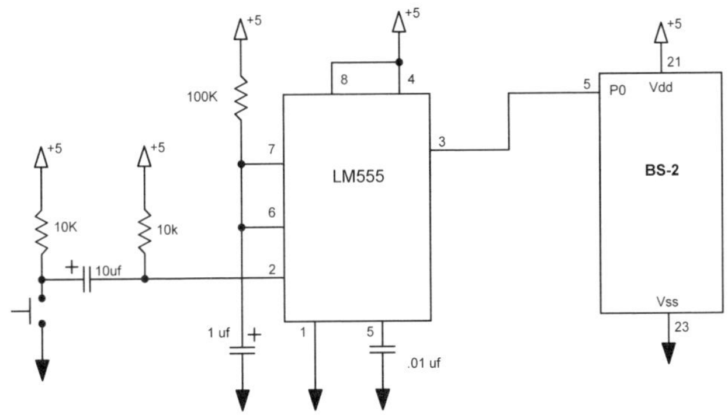


**Figure 3.5**
***A 555 timer switch de-bounce circuit***

**Code 3.5**

```
here:
if in0=1 then there
goto here
there:
debug "switch pushed"
goto here
```

Remember that since this is a "one-shot" circuit, once you push the switch it will generate a single output pulse – from low to high, and then back to low. It will not generate another pulse until the switch is released and pushed again. Therefore, this circuit cannot be used to detect the current state of the switch.

If you need to monitor the status of a switch (for example, whether or not a door is open or closed), then you would need to use a circuit such as shown in Figures 3.1 or 3.2.

If the switch needs to toggle between two states, you can use the circuit shown in Figure 3.6. It also provides "bounce-less" transitions to both states.

This circuit uses a 74LS74 "flip flop" and will toggle states each time the switch is pushed. In other words, push and release the switch once and P0 will detect a "1." Push and release the switch again and P0 will see a "0." This effectively creates a "push-on, push-off" switch circuit.

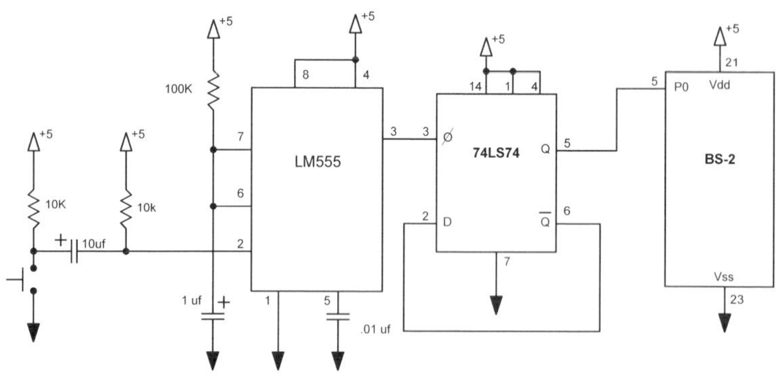


**Figure 3.6**
***A "push-on, push-off" bounce-free switch***

**Code 3.6**

```
here:
if in0=0 then I_am_low
debug "P0 detects a HIGH signal"
debug cr
goto here
i_am_low:
debug "P0 detects a LOW signal"
debug cr
goto here
```

If you need to connect a switch that has a voltage exceeding the safe operating limits for the microcontroller (usually no more than 5 volts), use the circuit in Fig. 3.7. The zener diode will snub any excessive voltage, and prevent damage to the I/O port.

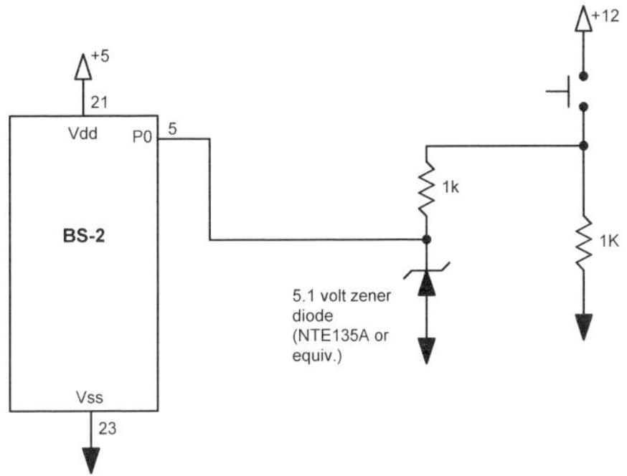


**Figure 3.7**

***The zener diode clips the input voltage down to a safe level***

**Code 3.7**

```
here:
if in0=1 then there
goto here
there:
debug "12 volt switch pushed, I detect 5 volts on P0"
debug cr
goto here
```

If you need to detect a voltage lower than the threshold voltage of the micon, you can use the circuit shown in Fig. 3.8. The values of the resistors connected to pin 4 (of the LM339) can be adjusted to accommodate the required "threshold" level (of the signal). Figure 3.8 has values calculated for a 0.8 "high" signal. This circuit is called a "comparator," and will produce a "high" signal any time the input voltage exceeds the voltage set on pin 4.

The LM339 has an "open collector" output, so to get a valid logic level we need to "pull it up." That's the purpose of the 15K resistor on the output of the LM339.

The 1 Meg ohm resistor provides hysteresis, eliminating possible oscillation during output transitions.

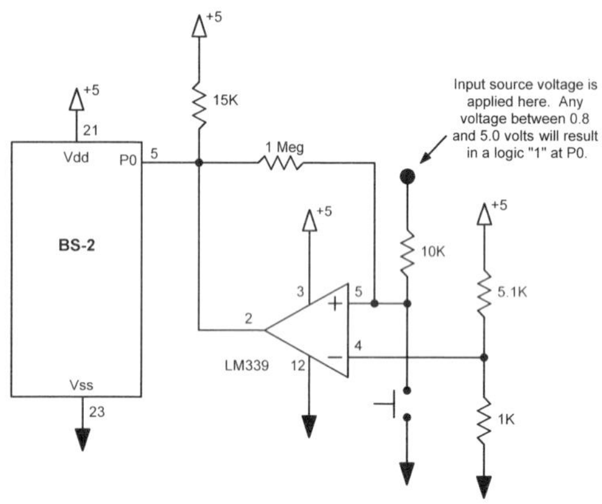


**Figure 3.8**
***Reducing the threshold level for a logic "high"***

**Code 3.8**

```
here:
if in0=1 then there
debug "Voltage is less than 0.8 volts"
debug cr
goto here
there:
```

```
debug "Voltage input has exceeded 0.8 volts"
debug cr
goto here
```

If your project requires a microcontroller to “read” a binary value, check out Figure 3.9.

This switch is a rotary hexadecimal type (from Mouser Electronics). By rotating the switch through 16 distinct positions, the 4-bit output is the binary equivalent of the rotary switch setting. With the switch set to “0,” the output presented to the micon is “0000.” With the switch set to “8,” the output is “1000,” etc.

These switches are available in binary coded decimal (“BCD”) versions as well.

A switch circuit of this type might be useful when you need to set different parameters for your device, but don’t require the full functionality of a “keypad” interface.

Although you could accomplish the same result with a “4 gang” DIP switch, using a rotary type doesn’t require the “user” to know the binary number system.

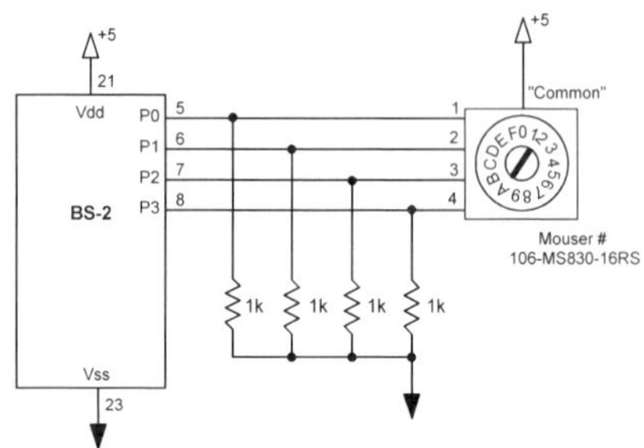


**Figure 3.9**
***A rotary hexadecimal switch input circuit***

**Code 3.9**

```
x var byte
here:
x=ina
debug ? x
goto here
```

# Photocell

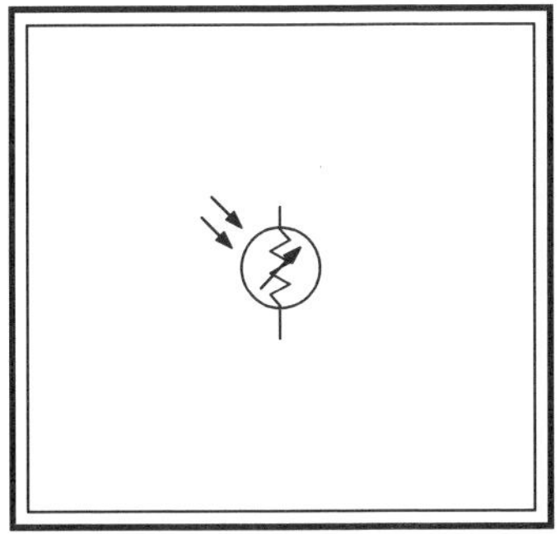

A photocell (also known as a "photo-resistor") is similar to a potentiometer, because it is a resistive device that can change its value.

However, instead of using the mechanical action of a rotating shaft to alter its value, a photocell's resistance changes based on the amount of light it "sees."

Photocells vary widely in their physical and electrical specifications; therefore, you may have to adjust some component values in the following diagrams to obtain proper circuit operation. Many photocells will exhibit a resistance of less than 1000 ohms when in bright sunlight, and multiple tens of thousands (or hundreds of thousands) of ohms when in total darkness.

Figure 3.10 illustrates the basic operation of a photocell. In this case, the photocell makes up one side of a voltage divider. As the amount of light striking the face of the unit changes, the voltage available at P0 changes.

In total darkness, the unit has a very high resistance, resulting in the analog voltage present on P0 to be pulled near to ground. The microcontroller detects this as logic "0." In bright light however, the resistance goes down, and P0 is pulled high (a logic 1).

While your code is running, cover and uncover the surface to the photocell and you'll see the Stamp displaying a 1 or 0, depending on whether it's light or dark.

By choosing the proper value resistor (or by substituting a potentiometer in its place), you can create a very inexpensive light detector, allowing your project to detect evening or morning.

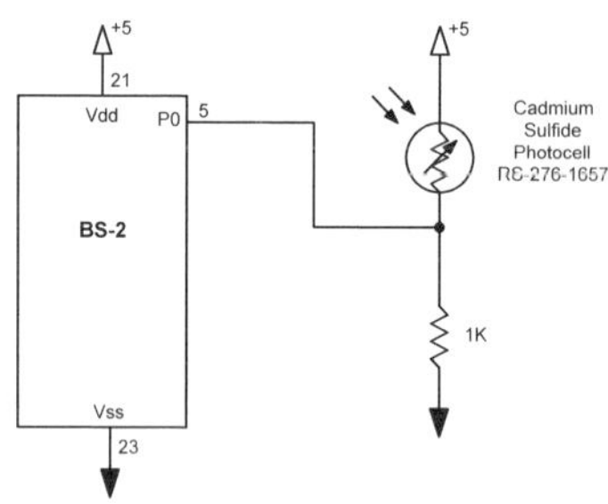


**Figure 3.10**
***A photocell used in a simple voltage divider network***

**Code 3.10**

```
x var bit
here:
x = in0
debug ? x
goto here
```

The Stamp has a nifty command called RCTIME that can be used to measure the intensity of the light striking the surface of the photocell. Figure 3.11 uses a capacitor and photocell to form a simple RC network.

By varying the amount of light falling on the cell, the Stamp can determine relative light intensity (without the need for an A/D converter).

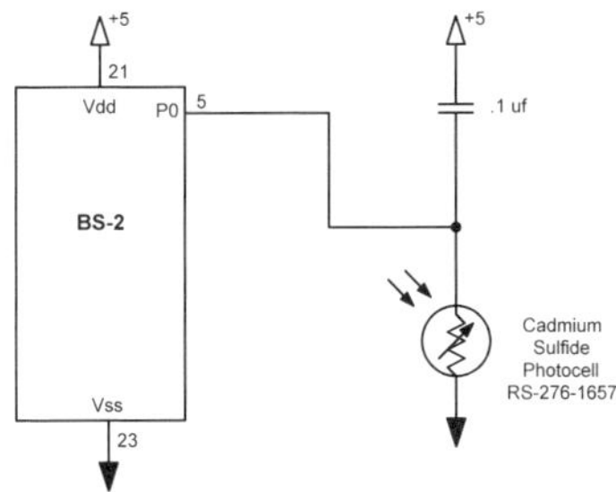


**Figure 3.11**
***The photocell used in an R/C time circuit***

**Code 3.11**

```
x var word
here:
high 0                  'discharge the capacitor
pause 1
rctime 0,1,x            'measure charge time
debug ? x
goto here
```

# Phototransistor

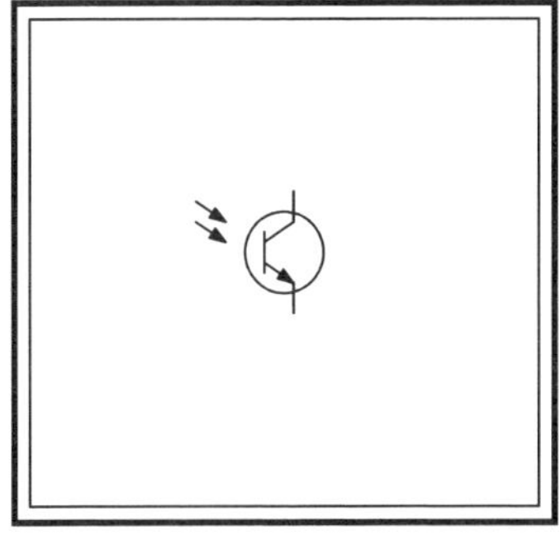

A phototransistor is a semi-conductor device that operates like a standard NPN switching transistor.

Instead of requiring a current to flow from the base to the emitter (to turn the transistor ON – see Chapter 4), a phototransistor conducts when light strikes the base. That's why there is a lens on the unit and, depending on the device, no electrical "Base" connection. See Chapter 4 for circuits that use standard NPN transistors.

Figure 3.12 illustrates the basic use for a phototransistor – that of an "electric eye." This circuit will not work over great distances, but by using some inexpensive lenses you can increase it's range dramatically.

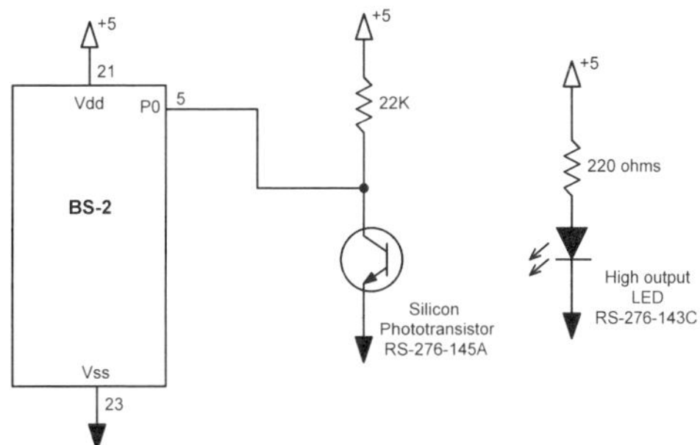


**Figure 3.12**
***A simple "electric-eye" circuit using a phototransistor***

**Code 3.12**

```
here:
if in0=0 then there
debug "It's dark"
debug cr
goto here
there:
debug "I see light"
debug cr
goto here
```

Phototransistors *can* be operated as linear devices; therefore, in “semi-light” conditions (when the phototransistor is biased in its *linear region*), the voltage input to the micon may oscillate and produce false signals. This is easily solved with the addition of a 74HC14 Schmitt trigger, as shown in Figure 3.13.

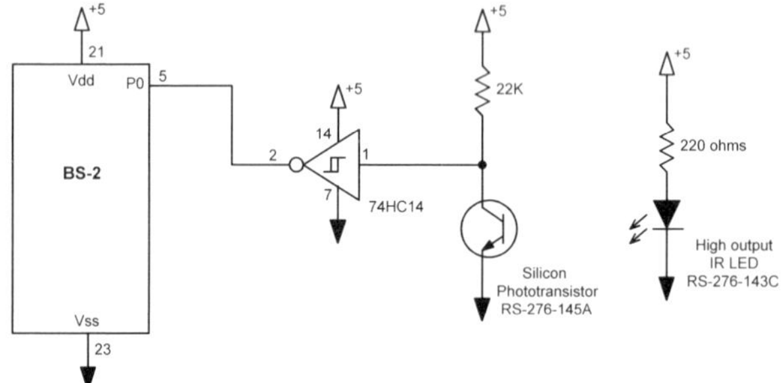


**Figure 3.13**
***“Cleaning up” the linear region of the phototransistor***

**Code 3.13**

```
here:
if in0=1 then there
debug "It's dark"
goto here
```

```
there:
debug "I see light"
goto here
```

Note also that the phototransistor shown is most sensitive to light in the infrared region; that's why we're using an infra-red (IR) LED.

We can't *directly* see whether or not the LED is working, but its operation can be verified by measuring the current flow through that portion of the circuit. Alternatively, you could use an IR light sensor like Radio Shack's #276-1099.

# Solar Cell

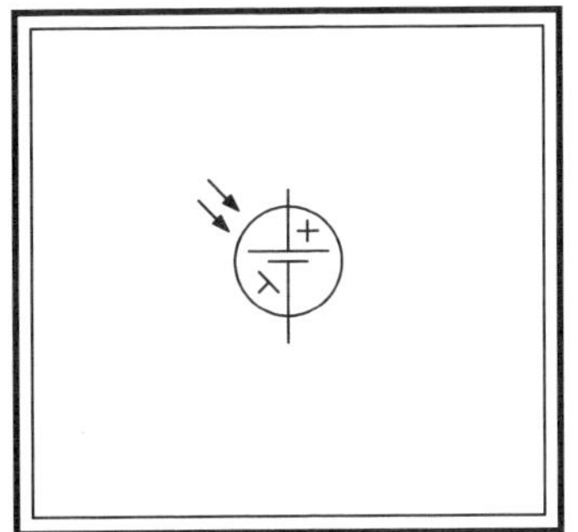

A solar cell generates a voltage when its surface is exposed to light. This voltage can be used to provide power to your project, eliminating the need for a battery or wall transformer.

A typical solar cell will produce approximately .3 to .5 volts – not enough to power a 5-volt microcontroller circuit.

However, just like batteries, you can connect solar cells together to obtain higher voltages.

Figure 3.14 shows 16 cells connected in series that will yield up to 8 volts DC when exposed to direct sunlight. By connecting this to a 5-volt regulator (or in the case of the Stamp, to Vin), your project will be able to operate whenever light is available. This is a great power source for applications such as remote data logging systems, etc.

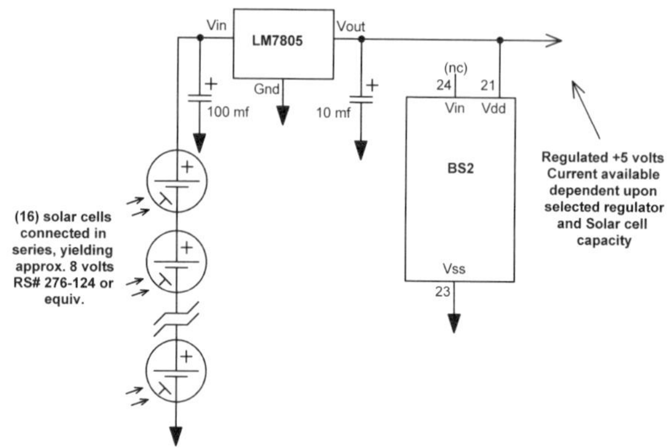


**Figure 3.14**

***Powering your microcontroller circuit with sunlight***

**Code 3.14**

"no code"

A solar cell can also be used as a sensitive light detection device. Figure 3.15 incorporates a single solar cell connected to the inputs of an LM324 operational amplifier. You can change the gain of the circuit by adjusting the potentiometer. This circuit might be used as a receiver for a digital light wave data transmission system.

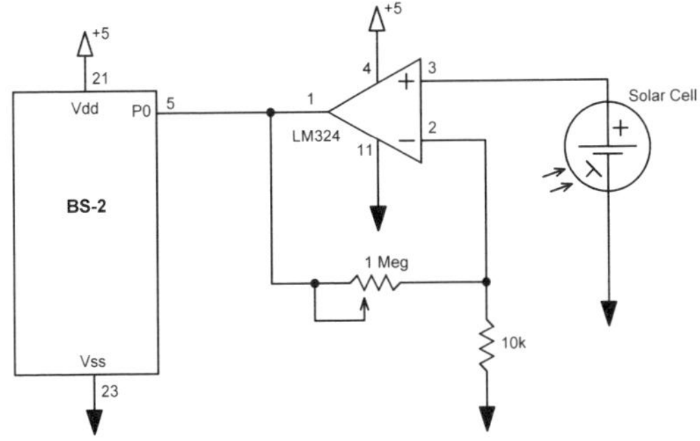


**Figure 3.15**
***A sensitive light detection circuit***

**Code 3.15**

```
x var bit
here:
x=in0
debug ? x
goto here
```

Another application for solar cells is to measure specific light levels. Utilizing an inexpensive A/D converter you can create a very sensitive digital light level measuring device, as shown in Figure 3.16. The ratio of the two resistors connected to pin 5 (on the AD0831) sets the span over which the A/D converter operates - see Chapter 5 for more information on A/D converters. This circuit is set for a full-scale reading of .5 volts.

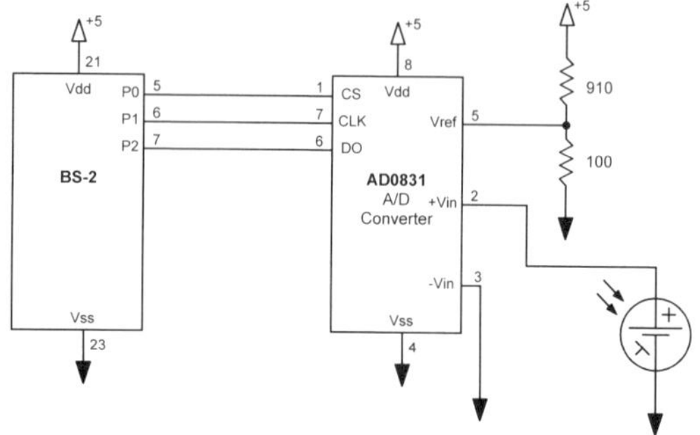


**Figure 3.16**
***Converting the analog voltage produced by a solar cell to a digital value***

**Code 3.16**

```
x var byte
y var byte
here:
low 0             'enable the 0831
pulsout 1,1       'send the setup clock pulse
x=0               'set x to 0
for y=1 to 8      'loop 8 times to get 8 data bits
pulsout 1,1       'send a clock pulse
x=x*2             'shift the bits left
x=x+in2           'add x to the next incoming bit
next
high 0            'disable the 0831
debug ? x         'display the result
goto here         'go do it again
```

# Thermistor

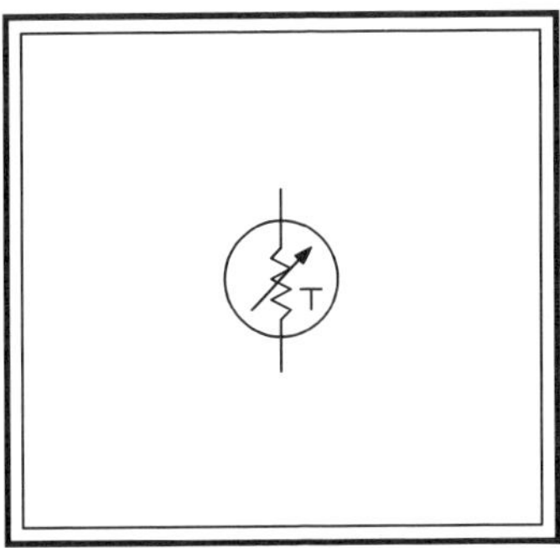


A thermistor is a temperature dependent resistor. Its method of operation is quite similar to a photocell.

Rather than responding to changes in light however, temperature variations cause a thermistor's resistance to fluctuate. This change in resistance can be detected in several different ways.

Figure 3.17 shows a thermistor connected directly to an I/O pin of the Stamp. The circuit uses the same RC time principles we used in the Photocell circuits.

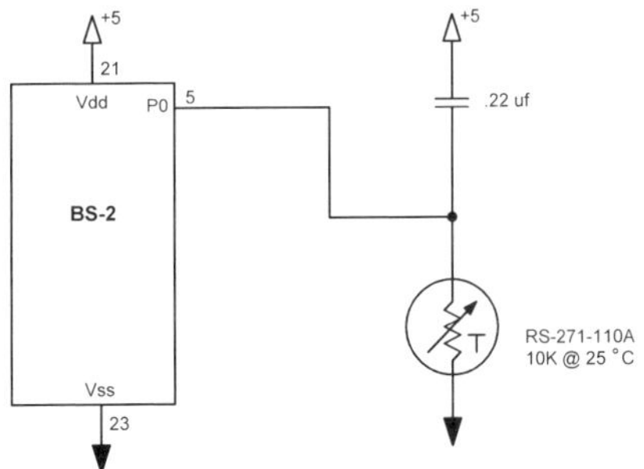


**Figure 3.17**
***A thermistor used in a simple R/C time circuit***

**Code 3.17**

```
x var word
here:
```

```
high 0
pause 1
rctime 0,1,x
debug ? x
goto here
```

A thermistor is an analog device. If your microcontroller doesn't have the "RCtime" command (built into the Stamp), but does have the ability to measure time between pulses, you could use the circuit shown in Figure 3.18.

Since a thermistor changes its resistance (due to temperature fluctuations), you could connect it to an LM555 timer circuit. In this circuit, the pulse rate of the LM555 changes in relation to temperature variations. The time period between pulses is then measured and a relative value is obtained. Depending on the temperature that you're measuring and the speed of your microcontroller, you may need to adjust the value of the capacitor connected to pin 2 on the LM555.

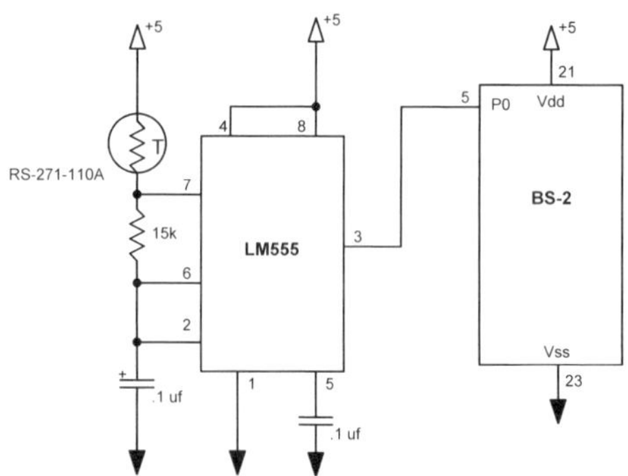


**Figure 3.18**
***Changing resistance in the thermistor alters the pulse width of the 555 timer***

**Code 3.18**

```
x var word
here:
pulsin 0,1,x
debug ? x
goto here
```

Alternatively, you can use a simple A/D converter chip and have the microcontroller simply read a binary data stream. See Figure 3.19.

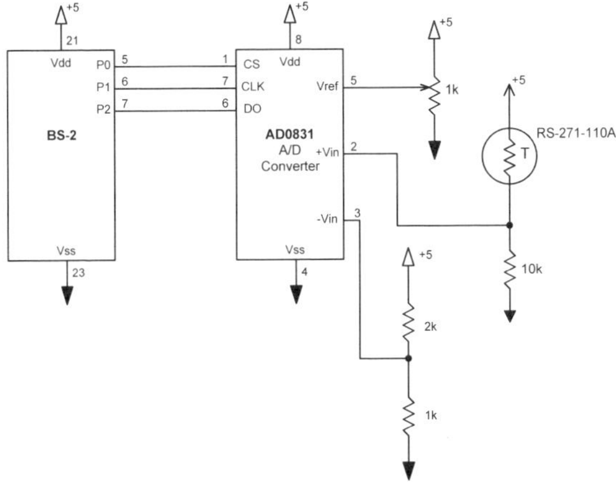


**Figure 3.19**

***A thermistor used in a voltage divider, and then converted to a digital value***

**Code 3.19**

```
x var byte
y var byte
here:
low 0                'enable the 0831
pulsout 1,1          'send the first setup clock pulse
x=0                  'set x to 0
for y=1 to 8         'loop 8 times to get 8 data bits
pulsout 1,1          'send a clock pulse
x=x*2                'shift the bits left
x=x+in2              'add x to the next incoming bit
next
high 0               'disable the 0831
debug ? x            'display the result
goto here            'go do it again
```

# Potentiometer

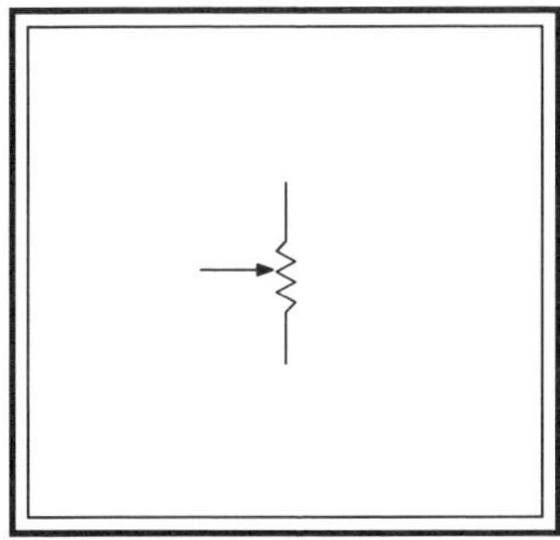

A potentiometer is a resistor that can change its value, usually through a mechanical action.

Physically, they come in two primary configurations: "rotary" and "slide." The electrical properties are identical, but like their names suggest, one rotates (like the volume control on a radio) and the other slides (like you would see on a recording studio mixer board).

Electrically, "pots" come in two versions as well: "linear" and "audio." Figure 3.20 illustrates the difference between these graphically. This is called the "taper," and is a function of how much the resistance changes in relation to the movement of the pot's wiper element.

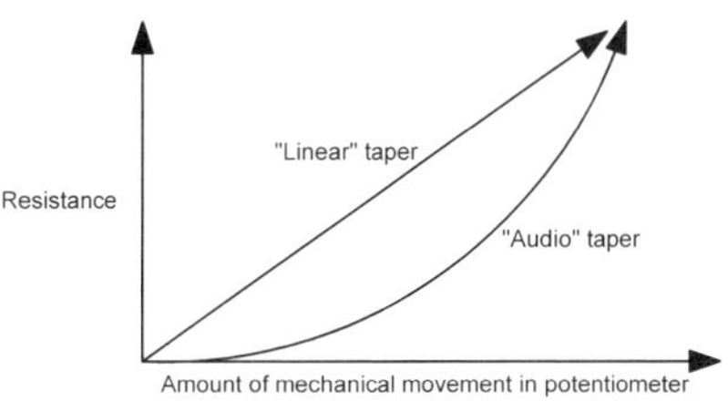


**Figure 3.20**

***The differences between linear and audio taper potentiometers***

**Code 3.20**

"no code"

You can see that the “linear” type changes resistance in direct proportion to the amount of movement in the wiper element.

The audio type changes resistance slowly at first. But then as you move the wiper, the “rate of change” continues to increase throughout its travel. The audio taper device is typically used in volume controls on radios, etc.

The circuit in Figure 3.21 uses a linear taper pot. However, you can use either type depending upon your application.

This circuit uses one of the built in commands of the Stamp, namely “RCtime.” As you run the program, you’ll notice that the value of “x” changes in direct proportion to the movement of the pot’s wiper. This can be used as a means to determine the relative position of the wiper arm, without using an A/D converter.

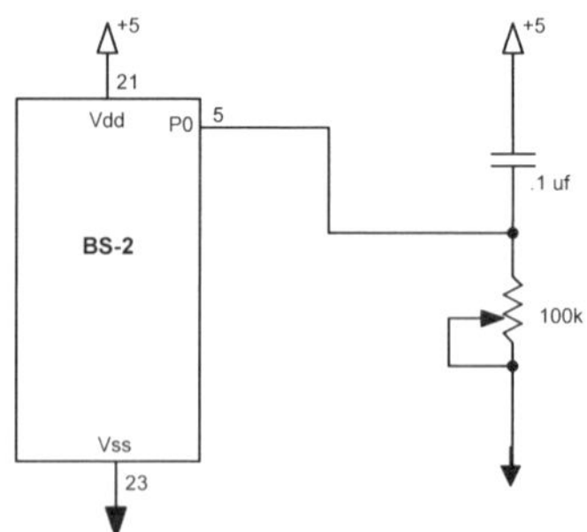


**Figure 3.21**
***A pot in an R/C time circuit***

**Code 3.21**

```
x var word
here:
high 0
pause 1
rctime 0,1,x
debug ? x
goto here
```

In some applications you may wish to detect the end of a mechanical movement, such as the closure of a door. Normally this could be accomplished with the use of a simple switch, but the positioning of the switch may be difficult in certain configurations.

For example, the easiest place to mount the switch is on the side of the door opposite the hinges (near the door knob). This location would provide you with the greatest amount of "resolution" for switch actuation. It would be much more difficult to position the switch near the hinges because of the arc or swing radius.

By using the circuit in Figure 3.22 and mounting the pot onto the hinge mechanism itself, you can not only detect door closure, but also eliminate the "click" of a switch.

This circuit takes advantage of the "logic detection level" on the I/O lines of a microcontroller. Typically, this is 1.4 volts for the detection of a logic "1."

The potentiometer is used as a simple voltage divider, and by careful orientation of the pot, the circuit can precisely detect when the door has completed its travel (either open or shut, depending upon your application).

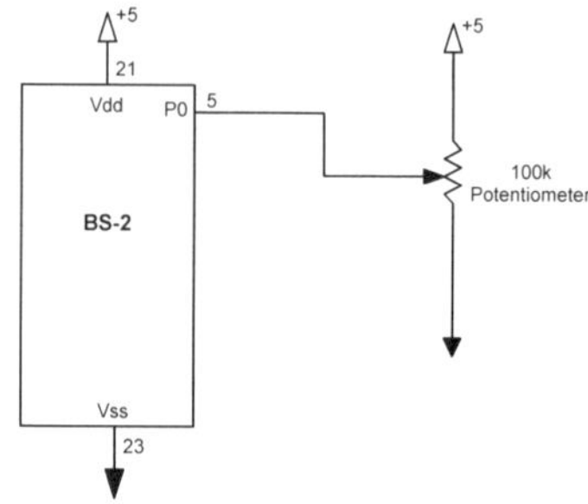


**Figure 3.22**

***A pot used as a voltage divider, taking advantage of the "threshold level" on the micon***

**Code 3.22**

```
x var bit
here:
x = in0
debug ? x
goto here
```

If you need to track the actual position of the wiper and your microcontroller does not have an "RCtime" type of command, you could use the circuit shown in Figure 3.23.

This circuit uses a simple serial A/D converter. The 8 bit converter results in 256 discrete values (or positions) for the wiper location.

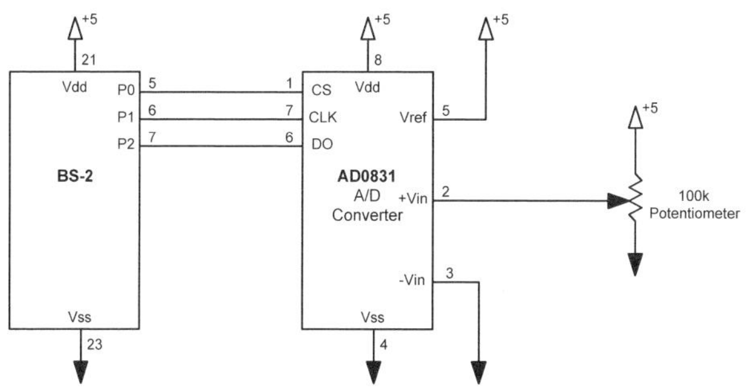


**Figure 3.23**
***The analog voltage from the pot is converted to a binary value***

**Code 3.23**

```
x var byte
y var byte
here:
low 0              'enable the 0831
pulsout 1,1        'send the first setup clock pulse
x=0                'set x to 0
for y=1 to 8       'loop 8 times to get 8 data bits
pulsout 1,1        'send a clock pulse
x=x*2              'shift the bits left
x=x+in2            'add the value of x to the incoming bit
next
high 0             'disable the 0831
debug ? x          'display the result
goto here          'go do it again
```

# Hall-effect Sensor

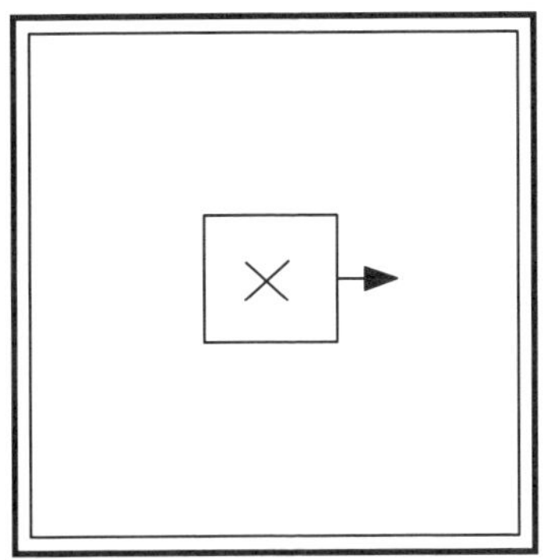

The "Hall-effect" was discovered (and subsequently named after) E.F. Hall in 1879.

Hall-effect sensors are semiconductor devices that change their output in relation to the influence of a magnetic field. The voltage output is directly proportional to the strength of the applied magnetic field. By definition, a Hall-effect sensor is a linear device (in its basic configuration).

Hall-effect sensors (like a phototransistor) can be used in "solid state switch" applications. Since there are no moving parts or contacts to wear out, they can also be used in "harsh environments."

Applications for Hall-effect ICs include use in ignition systems, speed controls, security systems, alignment controls, micrometers, mechanical limit switches, computers, printers, disk drives, keyboards, machine tools, key switches, and pushbutton switches. They are also used as tachometer pickups, current limit switches, position detectors, selector switches, current sensors, linear potentiometers, and brushless dc motor commutators (just to mention a few!).

Hall switches have an active area that is closer to one face of the package (the face with the lettering). To operate the switch, the magnetic flux lines must be perpendicular to this face of the package, and have the correct polarity. If an approaching south pole would cause a switching action, a north pole would have no effect. In practice, a close approach to the branded face of a Hall switch by the south pole of a small permanent magnet will cause the output transistor to turn "on."

Hall-effect sensors are available in two basic types: linear and digital.

A linear hall-effect sensor is ratio-metric. This means that the output voltage is (without any applied magnetic field) approximately ½ of the operating voltage. Therefore, if your circuit is operating on a 5-volt power supply, then the output will be approximately 2.5 volts. As the south pole of a magnet approaches the sensor, the output will increase. Conversely, as the north pole of a magnet approaches the sensor, the output voltage will decrease (below the 2.5 volt "quiescent" level).

Figure 3.24 uses the Allegro Microsystems A3518SUA. It is a linear sensor with a sensitivity of 2.5 mv/G. The AD0831 A/D converter measures the voltage output and delivers serial data to the microcontroller.

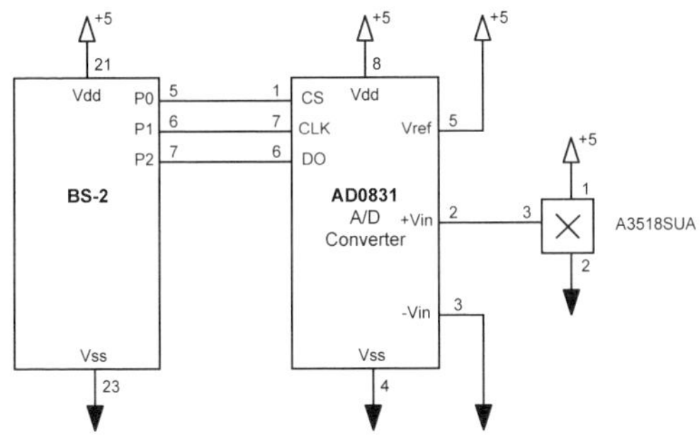


**Figure 3.24**
***A linear Hall-effect sensor, through an A/D converter***

**Code 3.24**

```
x var byte
y var byte
here:
low 0                    'enable the 0831
pulsout 1,1              'send the first setup clock pulse
x=0                      'set x to 0
for y=1 to 8             'loop 8 times to get 8 data bits
pulsout 1,1              'send a clock pulse
x=x*2                    'shift the bits left
x=x+in2                  'add x to the next in incoming bit
next
high 0                   'disable the 0831
debug ? x                'display the result
goto here                'go do it again
```

If your application requires a "switch operation" (using a digital type sensor), but you're not sure what value (measured in *gauss*) that your circuit should detect, you could use this circuit to determine the approximate value, and then select the appropriate digital sensor.

Digital devices are available in both uni-polar and bi-polar configurations.

The uni-polar device will switch "on" whenever there is a south pole field presence. When the field is removed from the detection area, the device switches "off." This device can be thought of as a "momentary" switch – in order for the output to remain "on," the (south) magnetic field must remain near the sensor.

In the bi-polar configuration, a south field presence will cause the switch to turn on. The output will remain on even after the removal of the south field. In order to turn the sensor off, you must introduce a north pole field. This type of device can be thought of as a "toggle" switch – one that requires an action in either direction to instigate a change in the output.

The digital version is essentially a linear Hall-effect sensor with a Schmitt-trigger contained within the same package. Digital sensors are available with different "trip point" settings. The circuit shown in Figure 3.25 uses the UGN3142 from Allegro Microsystems. This device is very sensitive – it can detect magnetic fields as low as 32 gauss. Other versions are available that will "switch" at levels of up to 400 gauss, or more.

Magnetic fields have two important characteristics—flux density and polarity (or orientation). In the absence of any magnetic field, most Hall-effect digital switches are designed to be OFF (open circuit at output). They will turn ON only if subjected to a magnetic field that has both sufficient density and the correct orientation.

A significant advantage inherent in the design of the Hall-effect sensor is that it has built-in hysteresis. This means that in order for the "switch action" to occur, you must exceed a certain level of field strength. Then to release the switch (turn OFF), the field strength must decrease substantially below the turn-ON level.

This offers built in noise immunity and results in a nice, clean switching action, free of "contact bounce." With a built-in Schmitt-trigger this sensor can be easily implemented in microcontroller applications.

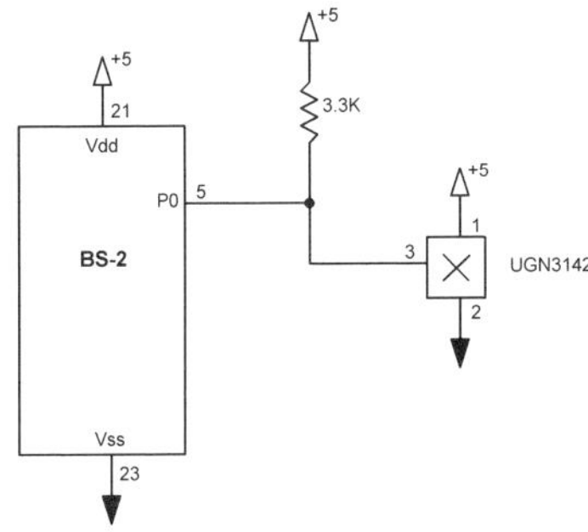


**Figure 3.25**
***A very sensitive digital Hall-effect sensor***

**Code 3.25**

```
x var bit
here:
x=in0
debug ? x
goto here
```

Figure 3.26 is an identical circuit to 3.25, but it incorporates the use of a bi-polar Hall-effect switch (UGN3132UA). This sensor turns ON with the approach of a south pole, and will remain ON (even when South is gone) until a North pole is detected. This circuit acts as a "toggle switch."

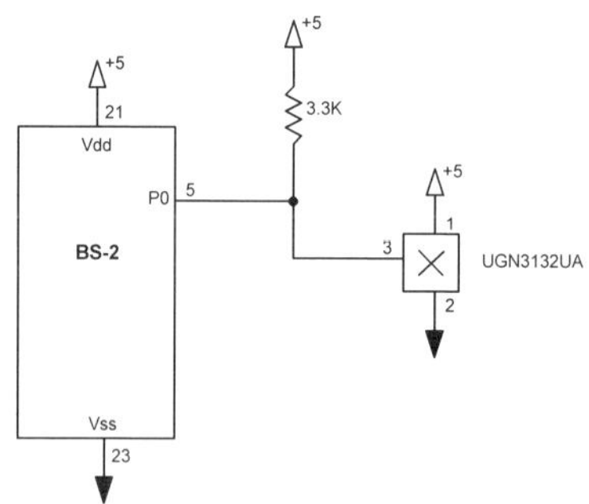


**Figure 3.26**
***A bipolar Hall-effect switch***

**Code 3.26**

```
x var bit
here:
x=in0
if x=0 then there
debug "I just saw a south pole of a magnet"
debug cr
goto here
there:
debug "That was a north pole"
debug cr
goto here
```

# Microphone

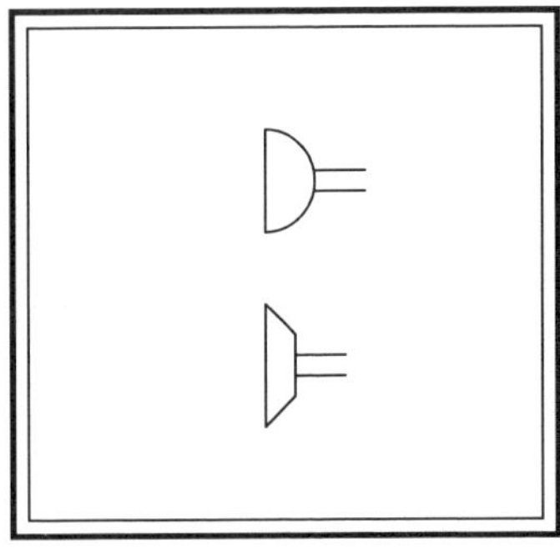

A speaker (see Chapter 4) receives a voltage from a circuit and directs it through a coil located in close proximity to a permanent magnet.

The resulting interaction of the magnetic fields causes a diaphragm to move. This diaphragm moves air, resulting in sound waves.

Conversely, sound waves can *cause* the movement of a diaphragm. This diaphragm (connected to a small coil of wire in close proximity to a permanent magnet), will induce a small voltage in that coil. This voltage can then be amplified and processed through your circuitry. This is the basic operation of a dynamic microphone.

Figure 3.27 uses an LM324 comparator IC that is connected to a dynamic microphone. When a certain sound level is reached (the gain of the LM324 can be adjusted by R1), the output voltage can be used to trigger an alarm. This circuit is extremely sensitive, and might make a great "intrusion detection" device.

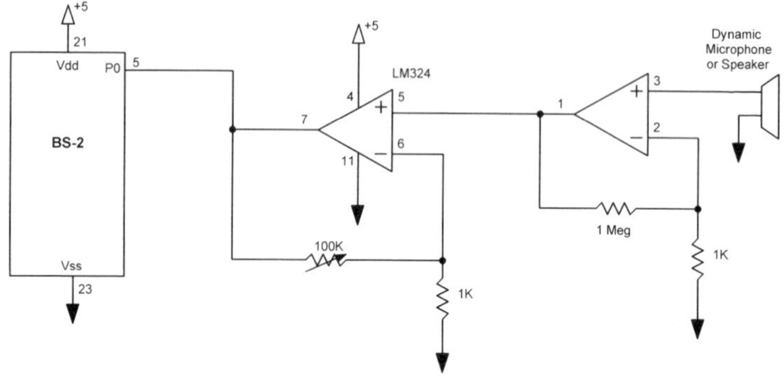


**Figure 3.27**
***A super-sensitive sound level detection device***

**Code 3.27**

```
x var bit
here:
x = in0
if x=1 then here
debug "I hear you!"
pause 100
debug cls
goto here
```

You can use a microcontroller to switch between different analog inputs, as shown in Figure 3.28. This circuit uses an analog multiplexer to select 1 of 8 different analog "channels" to be monitored.

Each one of these analog inputs could be connected to the output pin of an amplifier circuit we used in Figure 3.27. The microcontroller would then have up to eight "ears" – for sound level monitoring at different locations.

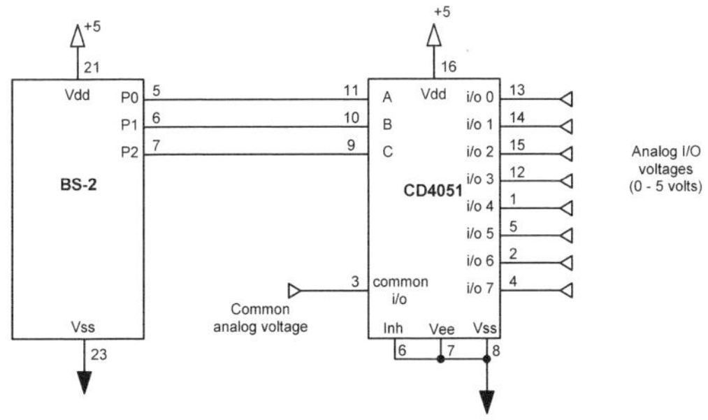


**Figure 3.28**

***Multiple analog voltages can be addressed through the 4051***

**Code 3.28**

```
x var byte
dira = 15
here:
for x= 0 to 7
outa = x
pause 500
next
goto here
```

# Humidity Sensor

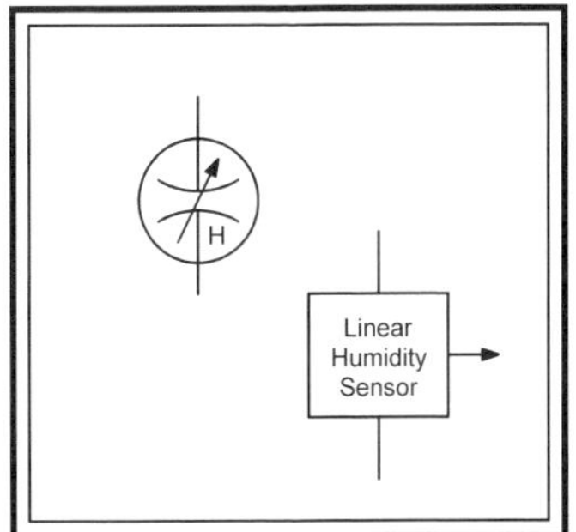


Humidity sensors attempt to measure the amount of moisture in the surrounding air. They can be used in many different types of applications such as refrigeration, drying, and meteorology.

Some consumer devices (TV's and VCR's) actually measure the Relative Humidity (RH) of the surrounding air, and will not allow the device to turn on if the RH is too high. This protects the electronic circuitry from damage.

Figure 3.29 uses an HIH-3605-A sensor made by Honeywell. This device has a linear voltage output from .8 to 3.9 volts over a span of 0-100% RH (at a temperature of 25 degrees C).

Ambient temperature does affect the accuracy of the sensor, and the above span will increase as temperature decreases. At 0 degrees C the span increases to 4.07 volts. Therefore, if you're measuring RH over a varying temperature range, you will need to compensate accordingly. See the following section for temperature measurement circuits.

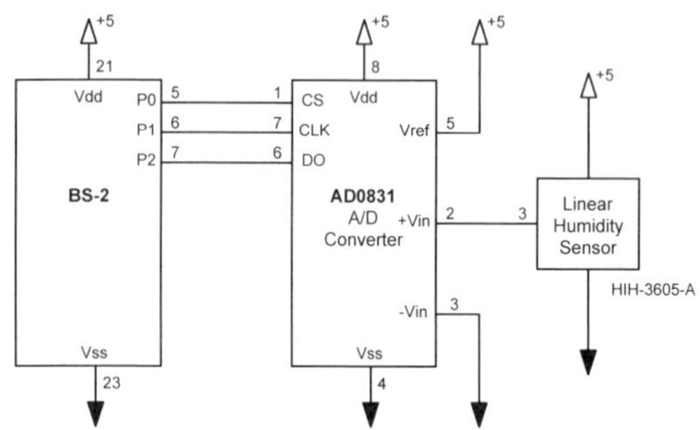


**Figure 3.29**

***The analog voltage produced by the humidity sensor is converted to a binary value by the A/D converter***

**Code 3.29**

```
x var byte
y var byte
here:
low 0                'enable the 0831
pulsout 1,1          'send the first setup clock pulse
x=0                  'set x to 0
for y=1 to 8         'loop 8 times to get 8 data bits
pulsout 1,1          'send a clock pulse
x=x*2                'shift the bits left
x=x+in2              'add x to the next in incoming bit
next
high 0               'disable the 0831
debug ? x            'display the result
goto here            'go do it again
```

# Linear Temperature Sensor

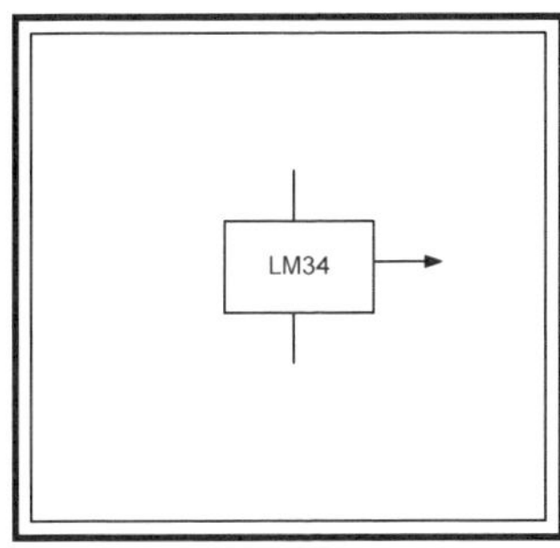


The National Semiconductor LM34 is a precision temperature sensor. It is calibrated to provide a linear output of 10 millivolts per degree Fahrenheit.

Figure 3.30 shows a typical application using a comparator to create a low temperature-warning device. Adjust the pot to an appropriate setpoint that, when exceeded, causes the output of the LM339 to go "high."

For example, if you want your microcontroller to be notified of "near freezing conditions", then set the voltage on pin 4 (of the LM339) to .36 volts. This establishes the temperature trip-point to 36 degrees Fahrenheit. As the ambient temperature drops below this point, the output of the comparator will go low, thus notifying the micon.

If you need to measure temperature in Centigrade, then use the LM35. Its operation is exactly the same, except the output is calibrated for 10 millivolts per degree C.

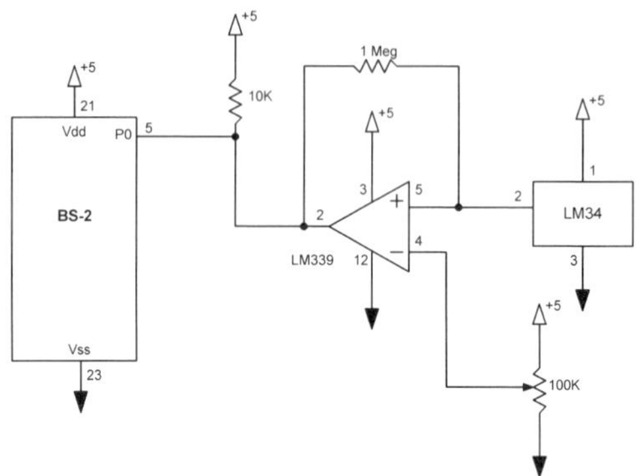


**Figure 3.30**
***A simple temperature "alarm" circuit***

**Code 3.30**

```
x var bit
here:
debug cls
x = in0
if x=1 then alarm
goto here
alarm:
debug "Temperature is near freezing"
pause 100
goto here
```

If your application requires the actual temperature to be known by the microcontroller, then use the circuit in Figure 3.31. The A/D converter can be calibrated for a limited span and a "zero offset" by adjusting the pots. For more information on the AD0831 (and A/D converters in general), see Chapter 5.

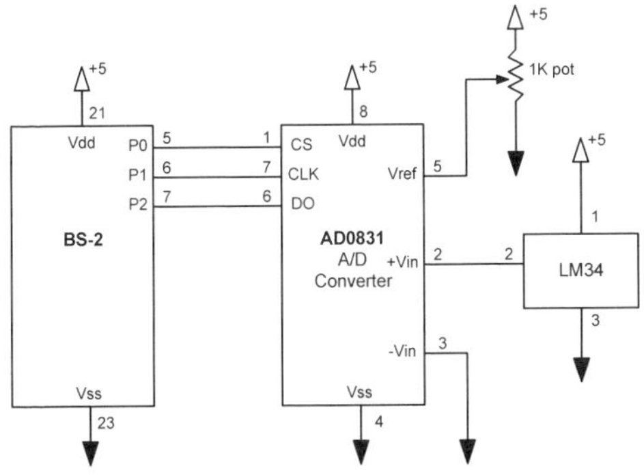


**Figure 3.31**
***Reading the digital value of the temperature sensor***

**Code 3.31**

```
x var byte
y var byte
here:
low 0                'enable the 0831
pulsout 1,1          'send the first setup clock pulse
x=0                  'set x to 0
for y=1 to 8         'loop 8 times
pulsout 1,1          'send a clock pulse
x=x*2                'shift the bits left
x=x+in2              'add x to the next in incoming bit
next
high 0               'disable the 0831
debug ? x            'display the result
goto here            'go do it again
```

# Optoisolator Input

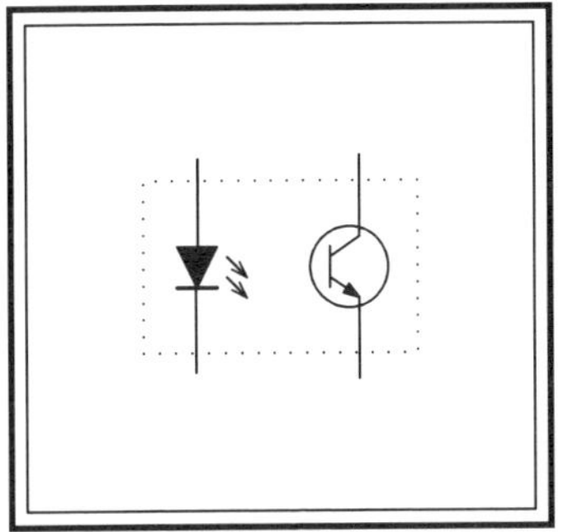

Optoisolators are used when you want to completely isolate two portions of a circuit. They are typically used in applications that may tend to generate voltage spikes or cause other types of circuit disruptions.

Essentially, an optoisolator is a device that can transmit information (usually digital) over a beam of light rather than through an electrical conductor. There are many different devices to choose from, depending upon your application requirements.

In this first circuit, there is no common electrical connection. The only way the information is getting into the microcontroller is via light from the LED encased in the optoisolator package.

Figure 3.32 uses a “PS2501,” made by NEC. There is really no electrical connection between the light source and the phototransistor, although they are contained within the same (4) pin dip package.

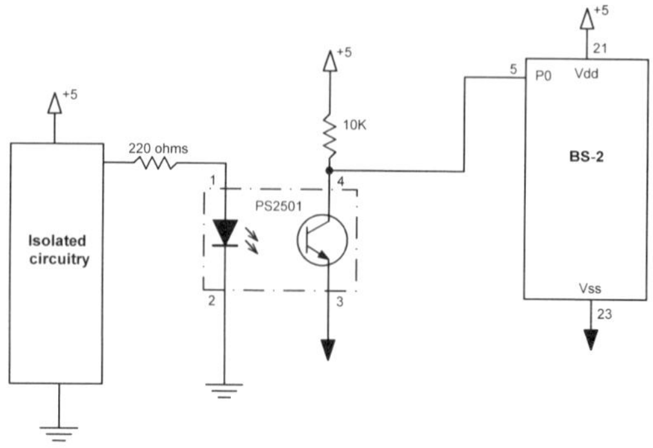


**Figure 3.32**
***The basic use of an optoisolator***

**Code 3.32**

```
x var bit
here:
x=in0
if x=0 then there
debug "Optoisolator is not seeing any light"
debug cr
goto here
there:
debug "The optoisolator is detecting a signal"
debug cr
goto here
```

The LED (contained within the optoisolator) does not have a current limiting resistor built into the device, therefore you must include one of an appropriate value in your circuit. The PS2501 specs call for no more than 80 ma through the LED.

If the input voltage source is not a digital signal, you may wish to add a Schmitt trigger as shown in Figure 3.33. The 74HC14 will clean up a noisy signal as its threshold level is reached, and produce sharp transitions for the micon.

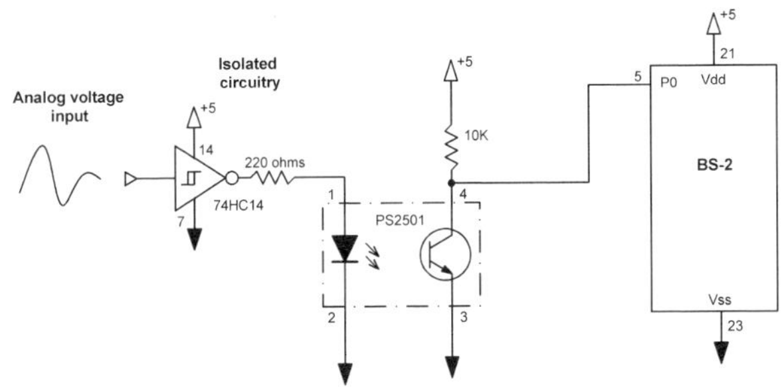


**Figure 3.33**
***An analog input signal can be "cleaned up" using a 74HC14***

**Code 3.33**

```
here:
if in0=0 then there
debug "It's dark in here"
debug cr
goto here
there:
debug "Gosh, it's bright in here"
debug cr
goto here
```

Although Figure 3.33 shows a common ground throughout the circuit, it doesn't need to be. It is a significant advantage to use light as a data transmission medium because it's possible to achieve complete electrical isolation (if required) in your application.

# 120 VAC Detection

The simplest (and safest) method to detect the presence of 120 VAC is to use a DC wall transformer as shown in Figure 3.34. The "wall wart" will step the voltage down to 12 Volts DC, where it is then further divided (using two resistors in series) to 5 VDC.

Not only do the resistors bring the voltage down to a safe level for the micon, but they also act as a "bleeder" circuit. Many wall transformers have built-in capacitors that can hold a charge far beyond the point where you lose power. The resistors present a load to the voltage source, draining any stored charge quickly, which ensures that the micon can quickly detect when power fails.

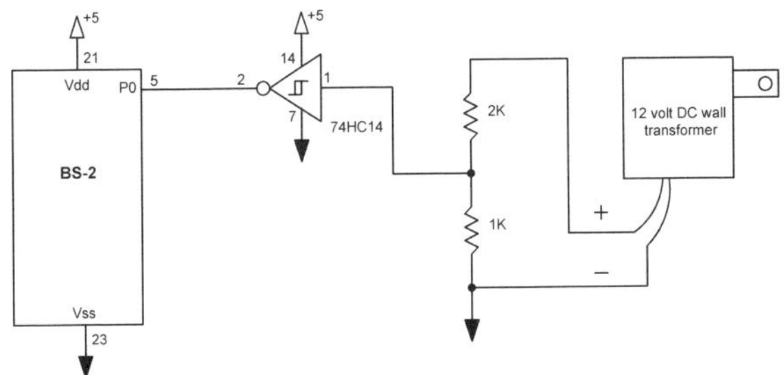


**Figure 3.34**
***Safe and simple detection of 120VAC***

**Code 3.34**

```
x var bit
here:
x=in0
if x=1 then there
debug "Power OK"
debug cr
goto here
there:
debug "Power Failure!"
debug cr
goto here
```

If your application does not allow the use of a wall transformer, then you can use the device shown in Figure 3.35.

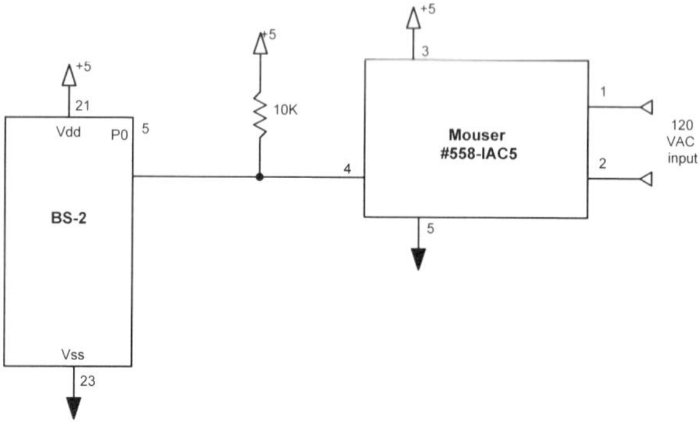


**Figure 3.35**
***Using an I/O module for 120VAC detection***

**Code 3.35**

```
here:
if in0=0 then there
debug "Power OK"
debug cr
goto here
there:
debug "Power Failure!"
debug cr
goto here
```

The Mouser #558-IAC5 is a solid state I/O module that is designed for connecting directly to 120 VAC. The output is an "open-collector," which explains the reason for the 10K pull-up resistor. In this circuit, we're using the same power source for the detection circuitry (contained within the module) as we are for the microcontroller itself.

This I/O module has all the built-in safety circuitry that virtually eliminates "accidents." But as with any electronic circuit, exercise caution when working with these higher voltages!

# Chapter 4
# Output

# Buffer

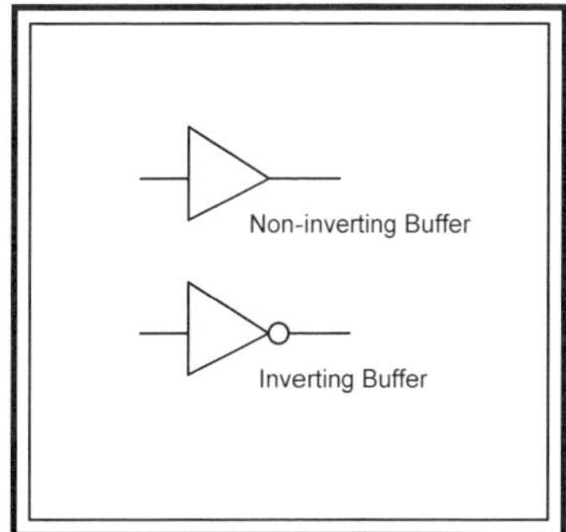


A buffer is a device or group of components that isolates two portions of a circuit. There are both digital and analog types of buffers, and each of these can be either "inverting" or "non-inverting."

Buffers are commonly used to increase the drive (amperage) capability of a signal.

For example, a typical microcontroller's I/O line might be limited to 20 milliamps. If you wanted to use that I/O signal to turn on a standard LED, it would probably work just fine. However, if you wish to drive a "high output" infrared LED, you might need as much as 100 milliamps (or more). Since the microcontroller could not safely deliver that amount of current, you will need to use a buffer circuit - perhaps something as simple as an NPN transistor.

Integrated circuit manufacturers have produced a large variety of buffers. Typically, there are several separate buffer circuits contained within the same IC package. One of the most popular buffers is the "7404" from the TTL series of logic devices. This device has been used in countless circuits and is available in many other "family types" such as HC, HCT, LS, ACT, etc.

For example, the circuit shown in Figure 4.1 specifies the "74HC04." The "HC" stands for "High-speed CMOS." Each "family" has slightly different electrical characteristics, but they all tend to have the same "pinout." However, be sure to double-check the appropriate data sheets of the devices you use in your project.

The 74HC04 comes in a 14 pin DIP format, and contains six independent buffer circuits.

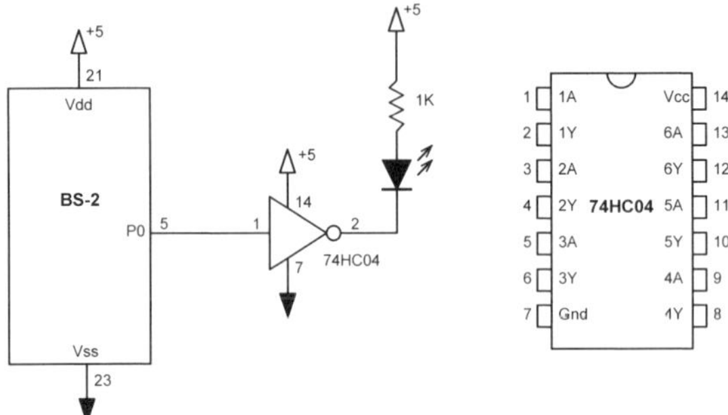


**Figure 4.1**
***The 74HC04 inverter IC***

**Code 4.1**

```
here:
high 0          'turn on the LED
pause 100       'wait for 1/10 second
low 0           'turn off the LED
pause 100       'wait again
goto here       'do it again
```

Figure 4.1 uses an inverting buffer. The signal that is applied to the input is inverted on the output. Sometimes this can be an advantage when writing your program. For example, in many "indicator" circuits, an LED is typically turned "on" by pulling its cathode to ground.

If the LED is connected directly to the I/O line on a micon, then you need to output a "0" (pulling the line low) in order

to turn on the LED. With an inverter in the circuit, a "high" signal from the micon is made low, thereby turning on the LED. This may seem more logical (turning something ON with a "high" from the micon rather than a "low"), and in more sophisticated programs, the easier it is to keep track of this sort of thing the better!

Figure 4.2 uses a 74HC05 inverting buffer. The output of this chip is "open-collector", therefore you can connect multiple outputs together as shown in the schematic. This can be useful in "multi-microcontroller" circuits, etc.

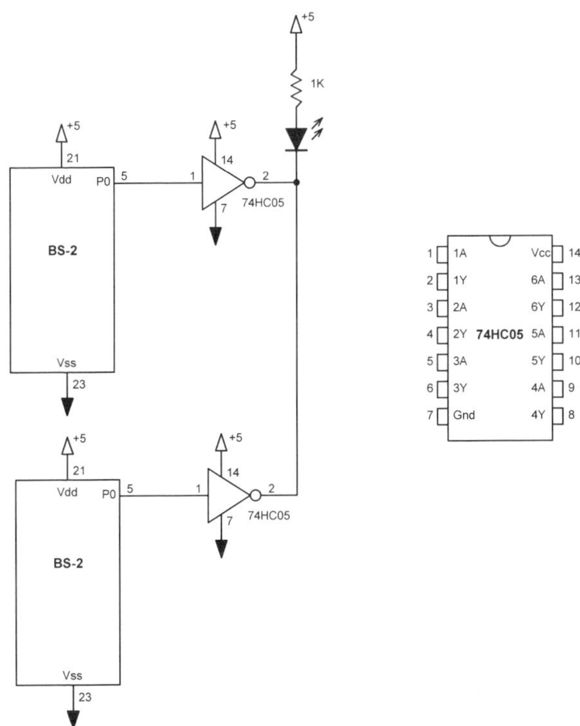


**Figure 4.2**
***The 74HC05 "open-collector" inverter IC***

**Code 4.2a**

```
here:             'load this program into the first stamp
high 0            'turn on the LED
pause 100         'wait for 1/10 second
low 0             'turn off the LED
pause 100         'wait again
goto here         'do it again
```

**Code 4.2b**

```
here:             'load this program into the other stamp
high 0            'turn on the LED
pause 135         'wait
low 0             'turn off the LED
pause 90          'wait again
goto here         'do it again
```

**Code 4.2c**

```
x var word        'load this program into both stamps
here:
for x=50 to 1000
high 0
pause x
low 0
pause 50
next
goto here
```

When a microcontroller's I/O line is programmed to be an "output," then by definition, it must be in either a high or low state.

By using a buffer such as the one shown in Figure 4.3, you can actually isolate the output from the circuit that it's driving. This is called a tri-state buffer, and will either be "high," "low," or "in a high impedance state." A high impedance state can be thought of as the buffer being able to "electrically disconnect" itself from the circuit.

This effectively means that no matter what level the I/O line is (on the micon), the tri-stated buffer is not allowing that signal to influence the other circuitry to which it's connected. This can be quite useful, especially in multi-microcontroller applications.

The 74HC241 contains 8 separate tri-state buffers within the same package. The buffers are separated into two groups of four. Buffers 1A1-1A4 are put into high impedance when $\overline{1OE}$ is pulled high. Buffers 2A1-2A4 are put into high impedance mode by causing 2OE to go low.

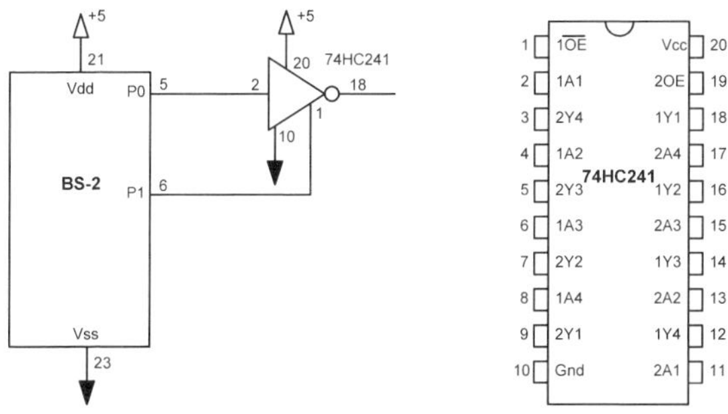


**Figure 4.3**

***A tri-state buffer can effectively isolate itself from the circuit***

**Code 4.3**

```
here:
low 1              'enable the '241 buffer chip
high 0             'turn on the output
pause 1000         'wait for a second
low 0              'turn off the output
pause 1000         'wait again
```

```
high 1                          'disable the '241
pause 1000
goto here                       'do it again
```

Try running Code 4.3 and touch a logic probe to pin 18 of the 74HC241. You'll see the output goes from high to low and then to "high impedance" – effectively no signal at all.

With some tricky programming you can maximize the use of the I/O lines by changing them from inputs to outputs (or vice versa), as required during the operation of your project.

The circuit in Figure 4.4 provides bi-directional buffer capability. Care must be exercised when using this type of circuit. Do not allow your program to make P0 an output at the same time that the 74HC245 is trying to deliver an input signal to the micon. The direction of the buffer is determined by the state of "DIR." Usually a separate I/O line from the micon controls this line (as demonstrated by P2 in Code 4.4).

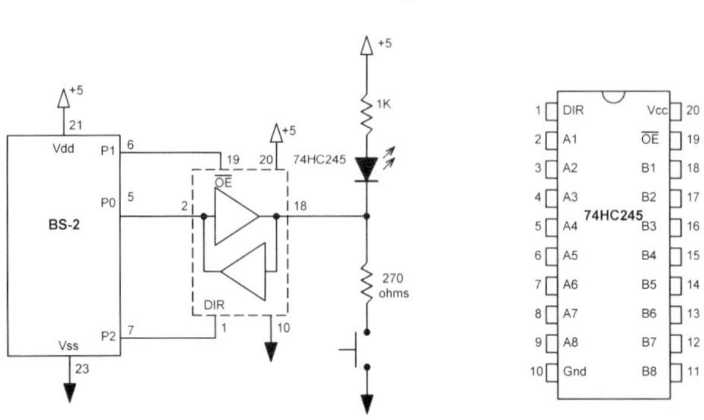


**Figure 4.4**
***A bi-directional buffer can maximize use of I/O lines.***

**Code 4.4**

```
x var byte
low 1                   'enable the '245

here:
debug "I'm blinking the LED now"
debug cr
for x = 1 to 50
high 2                  'cause direction to be "output"
high 0                  'turn on the LED
pause 50                'wait for 1/10 second
low 0                   'turn off the LED
pause 50                'wait again
next

low 2                   'make the direction an "input"
debug "I'm looking for an input"
debug cr

keeplooking:
if in0=0 then there     'wait for switch to be pushed
goto keeplooking

there:
debug "Thank you for pushing the switch"
debug cr
pause 1000
goto here
```

# Basic Gates

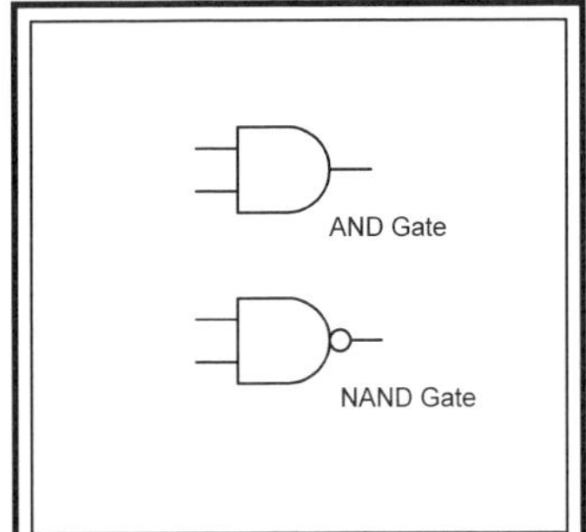


There are literally thousands of gates that make up a typical microcontroller. In all of their sophistication however, a micon may still need a few more for proper "project" operation.

A "gate" is a group of semiconductor components arranged in such a fashion as to perform some logical function.

The simplest is known as an AND gate. Like its name implies, an AND gate will only output a "high" signal when all of its inputs are high. AND gates come in many different configurations with 2, 3, or 4 (or more) inputs, and either non-inverting or inverting versions. The inverting version is called a NAND gate.

The 74HC08 IC contains four separate AND gates within the same package.

Figure 4.5 shows a typical application for a simple two input AND gate. In this multi-microcontroller system, the output will only be high when both P0 lines (from each micon) are high. Neither microcontroller is aware of what the other's status is, and only when both P0's are high does the output go "true" and turn on the LED.

A typical application for this kind of circuit might be as a "safety interlock" switch. Each microcontroller may be monitoring different parameters, but only when each has independently determined that conditions are met will the circuit produce a high output.

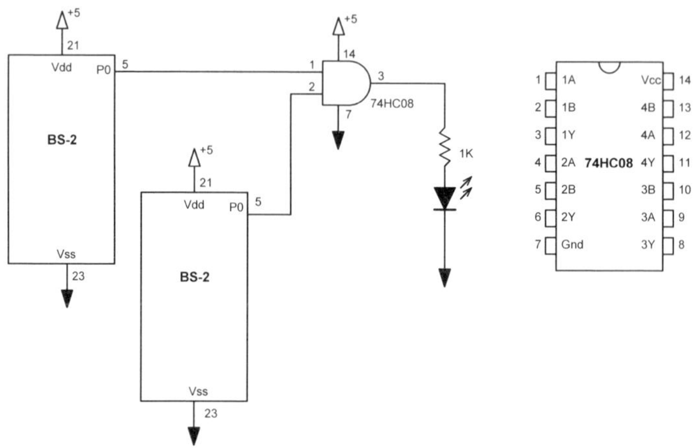


**Figure 4.5**
***The LED will turn on only when both inputs are high***

**Code 4.5**

```
x var word    'load program in both micons for random blinking
here:
for x=50 to 1000
high 0
pause x
low 0
pause 50
next
goto here
```

Figure 4.6 uses the same gate, but in a slightly different manner. In this case the microcontroller provides an

"enabling" function. If P0 is high, then each time the switch is pushed, a high is produced on pin 3 of the 74HC08.

Also, by connecting P1 to the output of the gate, the micon can monitor the output status of the circuit, thereby providing simple feedback.

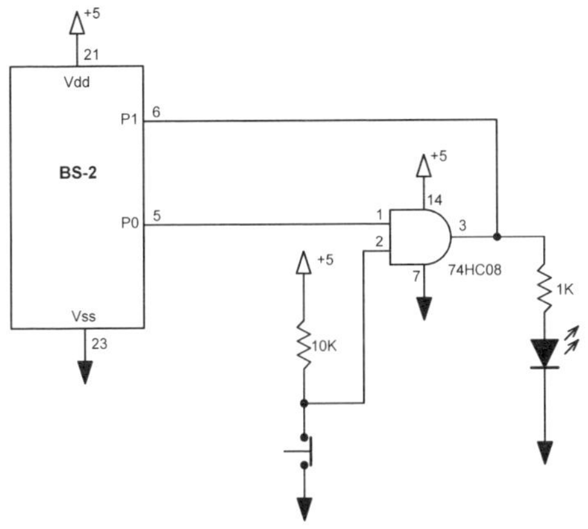


**Figure 4.6**
***The LED will light when you push the switch, but only if the microcontroller "allows" it***

**Code 4.6**

```
x var word
here:
low 0                         'disable the AND gate
debug "If you push the switch now, nothing will happen"
debug cr
pause 6000
high 0

debug "The LED will light if you push the switch now"
debug cr
lookagain:
```

```
if in1=1 then there               'look for the switch input
goto lookagain

there:
debug " You pushed the switch and the LED came on"
debug cr
pause 2000
goto here
```

Figure 4.7 is nearly the same circuit, but uses a 74HC32 OR gate instead. In this case, either P0 or the opening of the switch will cause the output of the gate to go high.

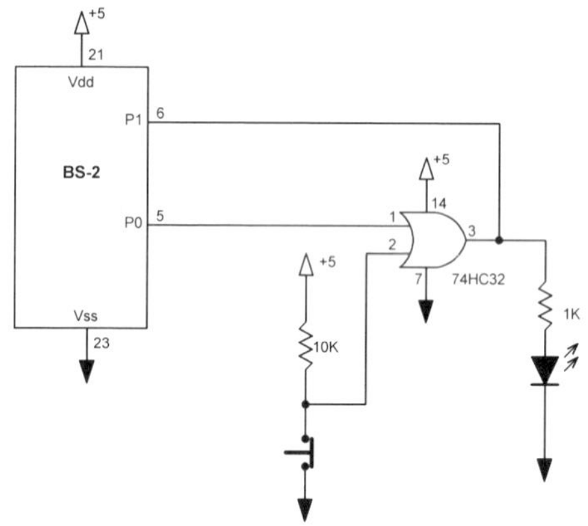


**Figure 4.7**
***Either the microcontroller or a pushed switch will turn on the LED***

**Code 4.7**

```
x var word
here:
high 0                            'disable the AND gate
debug "I have turned on the LED"
debug cr
```

```
pause 6000
low 0
debug "I turned off the LED, but you can turn it on by pushing
the switch"
debug cr
lookagain:
if in1=1 then there                'look for the switch input
goto lookagain

there:
debug "I noticed that you pushed the switch and the LED
came on"
debug cr
pause 2000
goto here
```

# LED

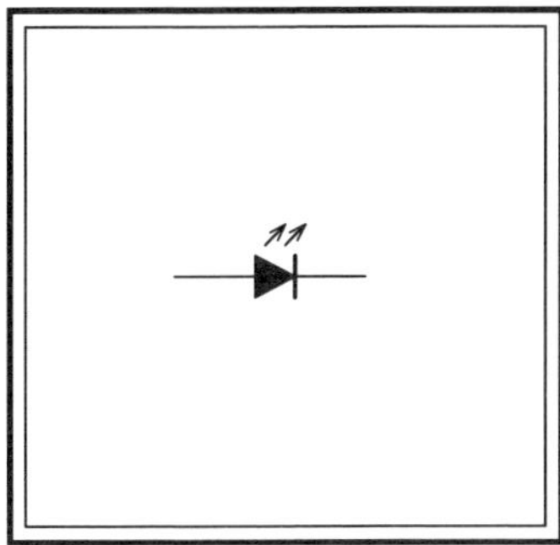

An LED is a semiconductor device. It is a diode that emits energy in the form of light.

The most common devices produce red light, but LED's are available in yellow, green and blue as well. High output infra-red units are also available.

LED's have a threshold voltage that must be exceeded before conduction will occur. This is typically around two volts, but can be higher depending upon the specific device. Once conduction occurs, as in a "typical" rectifier diode, there is very little resistance to current flow. Therefore, a current limiting resistor should always be included in the circuit to protect not only the LED, but the driver circuitry as well.

Because an LED will only conduct in one direction (like a standard diode), you must observe polarity when designing your circuit. Many microcontrollers can source (high) or sink (low) enough current to turn on an LED in either direction. In the case of the Stamp, you should limit this current to about 20 milliamps.

Figure 4.8 is a circuit that will turn on the LED when P0 goes low. The cathode of the LED is "sunk" to ground through the I/O line (as indicated by the arrows).

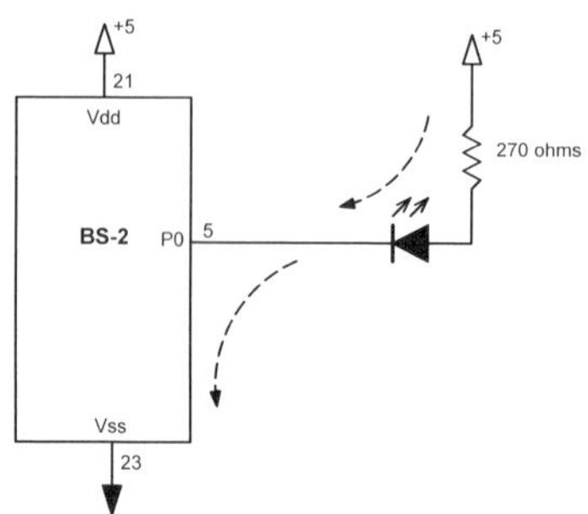


**Figure 4.8**
***The LED is "sunk" to ground through the microcontroller***

**Code 4.8**

```
low 0
stop
```

Figure 4.9 depicts a microcontroller "sourcing" current from the positive supply voltage thru the LED to ground. Some microcontrollers may not have enough "sourcing" capability to operate LED's in this manner, so be sure to check the specifications of the device you're using.

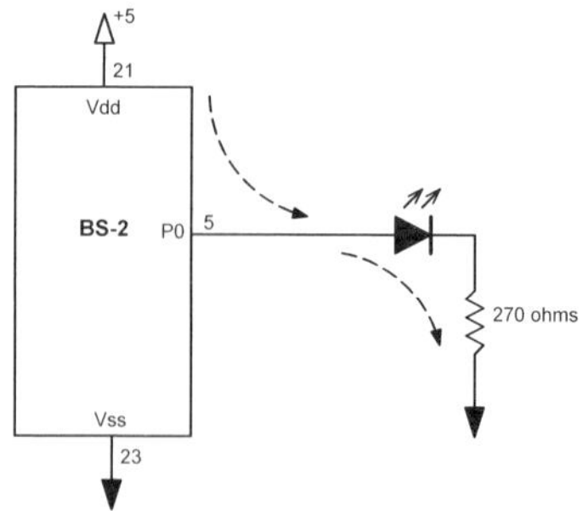


**Figure 4.9**
***The microcontroller "sources" current to the LED***

**Code 4.9**

```
high 0
stop
```

Radio Shack sells a high-output infrared LED that can typically operate at about 100 milliamps. Since this is well beyond the ability of most micons, a simple bipolar transistor or MOSFET buffer circuit, such as shown in Figure 4.10, is appropriate. In fact, because of the high current capabilities of the IRF511, you could actually connect several dozen LED's to the same circuit (all operating from the same MOSFET), resulting in a very bright infrared light source.

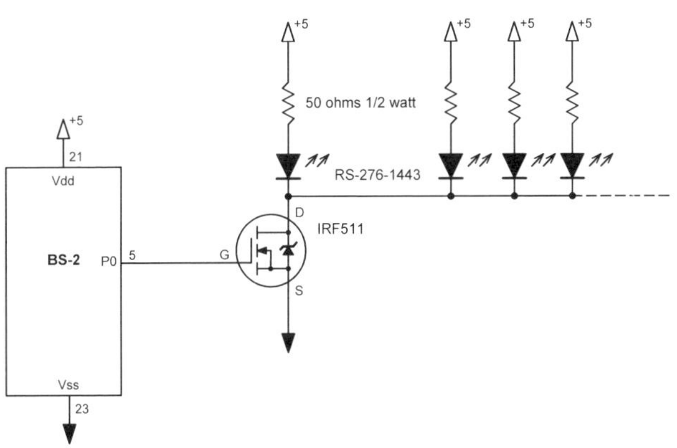


**Figure 4.10**
***A MOSFET can turn on a "heavy load"***

**Code 4.10**

```
here:
high 0
```

```
pause 100
low 0
pause 100
goto here
```

As shown in Code 4.10, a microcontroller can easily blink an LED on and off. However, since microcontrollers are not "multi-tasking" (able to do things *seemingly* simultaneously), it's sometimes difficult to blink an LED and still tend to other circuit requirements.

A blinking LED might be a good visual indicator, but while the LED is blinking it may take up precious processing time.

The circuit in Figure 4.11 eliminates the need for "creative programming" and solves the blinking LED problem with hardware. The LM3909 is designed specifically for this purpose. By turning on power to the LM3909 (via the transistor), the circuit will blink the LED at a rate determined by the value of the capacitor attached to pin 2.

With this circuit, all your program needs to do is set the output of P0 high and then go on about its business. With the circuit enabled, the LED blinks with no program intervention. Of course, a low on P0 turns off the blinker.

You can increase the blink rate by decreasing the value of the capacitor.

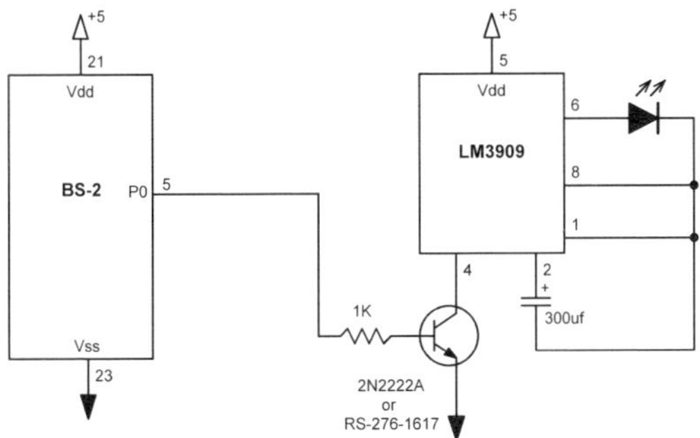


**Figure 4.11**
***The flasher circuit is enabled by the microcontroller***

**Code 4.11**

```
high 0
debug "The LED circuit is blinking by itself."
stop
```

# Incandescent Lamp

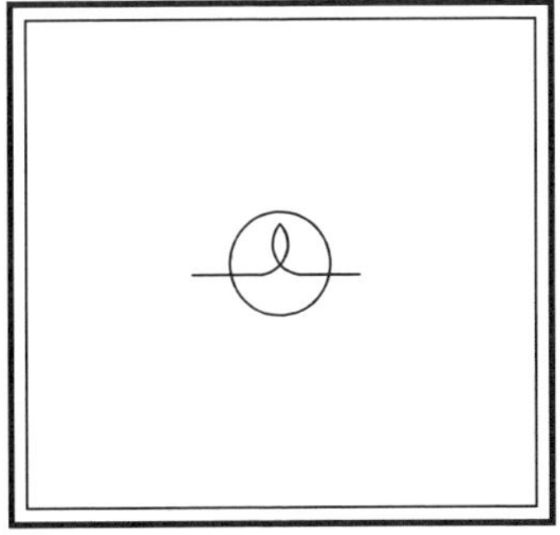

With the exception of very low intensity devices, most incandescent lamps draw far more current than can be safely supplied by a microcontroller. Therefore, you must use some sort of buffer or driver circuit, as shown in Figure 4.12.

This circuit uses an inexpensive MOSFET transistor that is capable of delivering several amps of current to the load.

When P0 is driven high, the transistor is turned "on" thereby allowing current to flow through the lamp to ground.

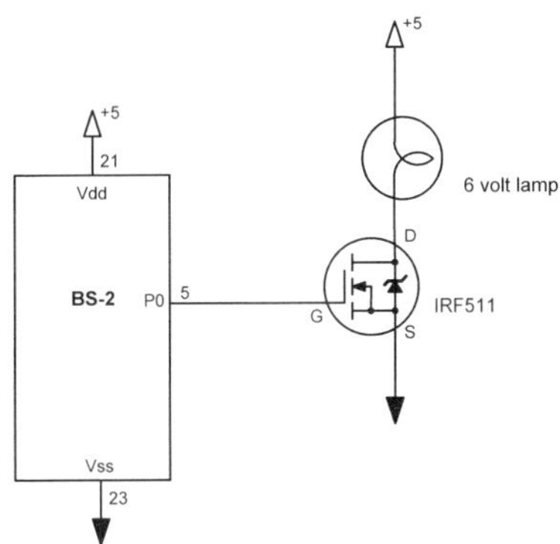


**Figure 4.12**
***A simple MOSFET lamp driver***

**Code 4.12**

```
here:
high 0
pause 100
low 0
pause 100
goto here
```

Figure 4.13 uses a 555 timer circuit to blink a low voltage lamp. In this manner, the microcontroller can blink a light without using up processing time or program space.

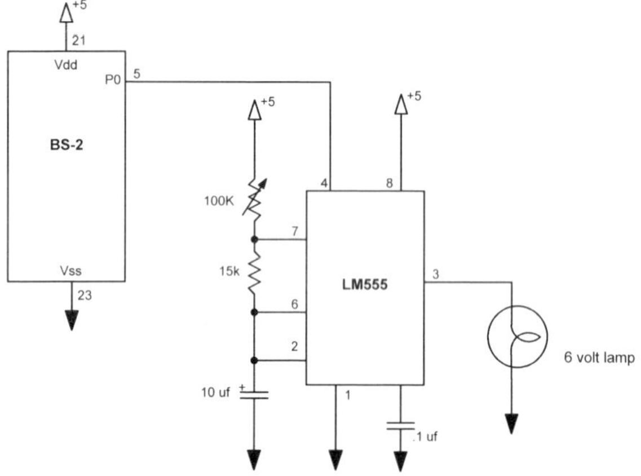


**Figure 4.13**
***The 555 timer blinks the lamp, but only if "enabled" by the microcontroller***

**Code 4.13**

```
here:
high 0          'enable the flasher circuit
pause 2000      'let it operate for 2 seconds
low 0           'disable the flasher
pause 4000      'for 4 seconds
goto here
```

Figure 4.14 uses a solid-state relay (SSR) to control a high voltage (typically 60 watt) lamp.

Many different models of SSR's can be used in this type of application. Most will require a relatively low DC voltage to turn on, and can be driven either high, as shown in Figure 4.14, or low (see Figure 4.15).

Be sure to use caution when working with line voltages, and follow any instructions or data sheets that came with your choice of SSR.

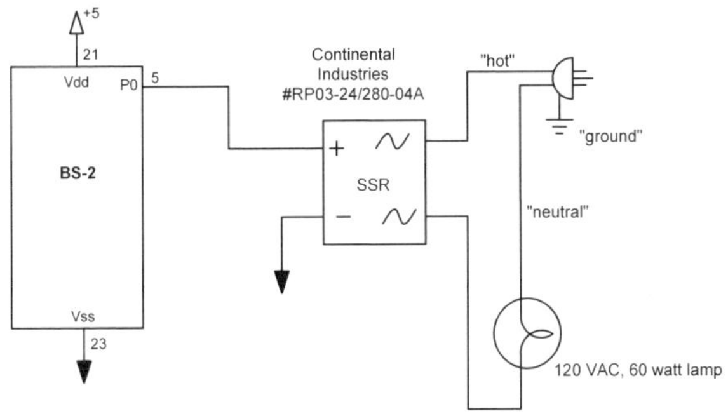


**Figure 4.14**
***Turning on a 60 watt light bulb by presenting a "high" signal***

**Code 4.14a**

```
high 0
stop
```

**Code 4.14b**

```
here:
low 0
pause 1000
```

```
high 0
pause 1000
goto here
```

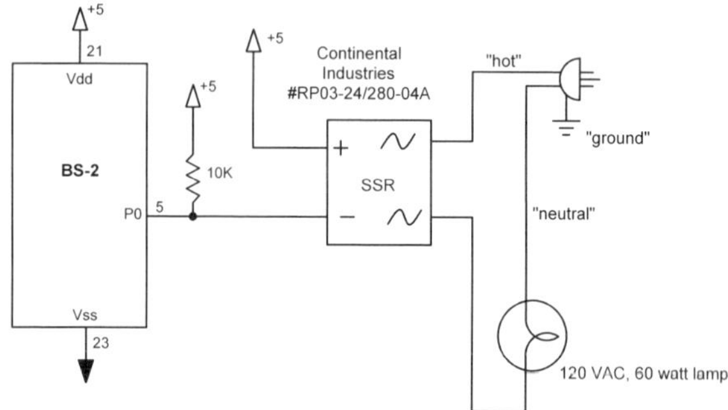


**Figure 4.15**

***Turning on a 60 watt light bulb by "sinking" to ground***

**Code 4.15**

```
low 0
stop
```

# Bipolar Transistor

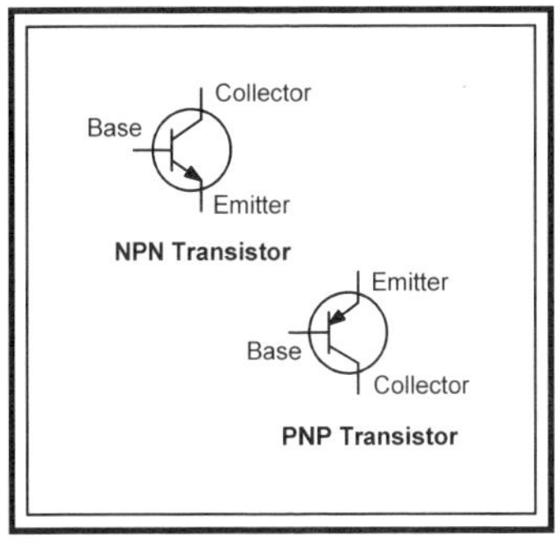


Bipolar transistors are available in two basic types: "NPN" and "PNP." Generally speaking, you can think of the NPN type as *usually* connected to the ground or "N"egative side of your circuit. Conversely, the PNP type is usually connected to the "P"ositive side.

Bipolar transistors have three connections. These are known as the "base," "emitter," and "collector."

Bipolar Junction Transistors (BJT's) are current actuated devices. This means that in order for a current to flow from the collector to the emitter (on an NPN type), you must first cause a (much smaller) current to flow from the base to the emitter. When this current flow is present, the transistor is "biased" or switched "on." When the transistor is "on," current will flow through the device from the collector to the emitter (effectively operating like a switch).

The amount of current necessary to turn on a transistor (such as the one used below) is quite low – far less than the capacity of a typical microcontroller. This makes interfacing to a micon quite simple, as shown in Figure 4.16.

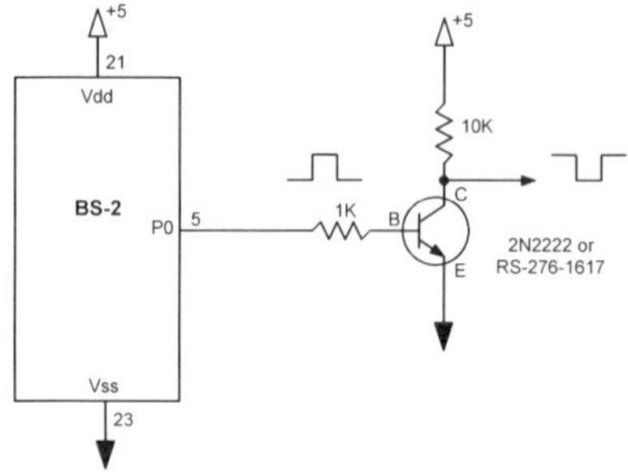


**Figure 4.16**
***A simple NPN bipolar transistor inverter***

**Code 4.16**

```
here:
low 0
pause 10
high 0
pause 10
goto here
```

This circuit can be thought of as a simple "inverting buffer," because the signal (shown as a positive going pulse in the schematic) is inverted as it goes "through" the transistor.

Figure 4.17 shows a typical use for an NPN transistor. When a "high" is presented on P0, a current is caused to flow through the base, thus turning on the transistor and effectively connecting the cathode (negative side) of the LED to ground. This is sometimes referred to as "sinking" current to ground.

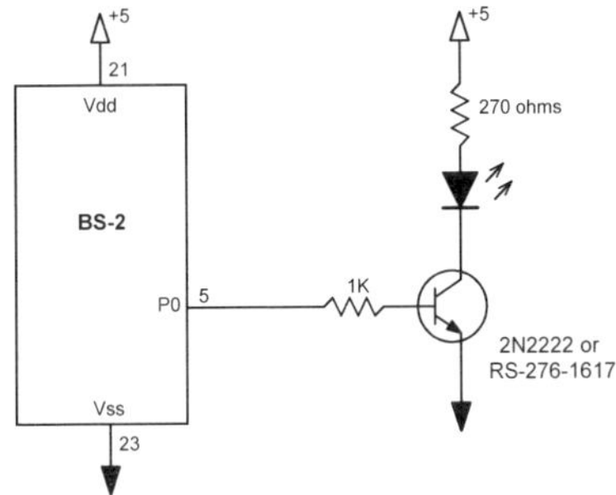


**Figure 4.17**

***When a "high" is present, the transistor turns on the LED by "sinking" it to ground***

**Code 4.17**

```
here:
low 0
pause 100
high 0
pause 100
goto here
```

Figure 4.18 shows the same application, but uses a PNP type of device. Notice that the emitter of the PNP transistor is connected to the "positive supply" side of the power source. Current is being "sourced" by the transistor to the load.

Although advancements in technology happen daily, NPN transistors are generally cheaper than "equivalent rated" PNP types because they are somewhat easier to make and are used more extensively in discrete applications.

Therefore, if your application can use either type of transistor, it is usually more efficient and cost-effective to use the NPN device.

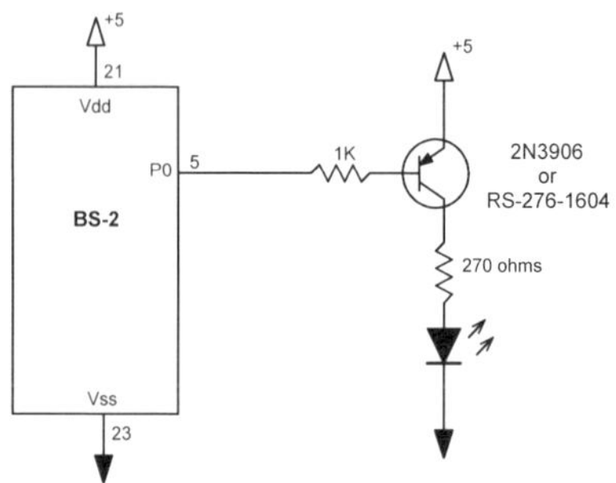


**Figure 4.18**
***A PNP transistor can "source" current to turn on a device***

**Code 4.18**

```
here:
low 0
pause 100
high 0
pause 100
goto here
```

Figure 4.19 depicts an "open-collector" application. In this circuit, either microcontroller can turn on the LED without worrying about "contention" with the other device. If your application requires, you can add as many additional micon-transistor combinations as necessary, by simply attaching each additional collector to the cathode of the LED.

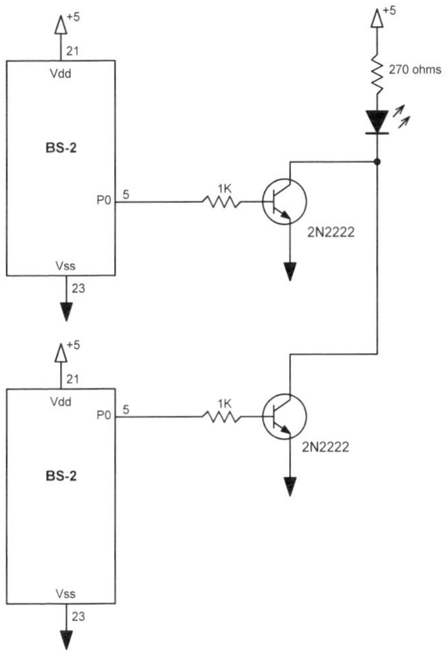


**Figure 4.19**

***NPN transistors can be used to create "open-collector" circuits***

**Code 4.19**

```
here:
low 0
pause 100
high 0
pause 100
goto here
```

# MOSFET Transistor

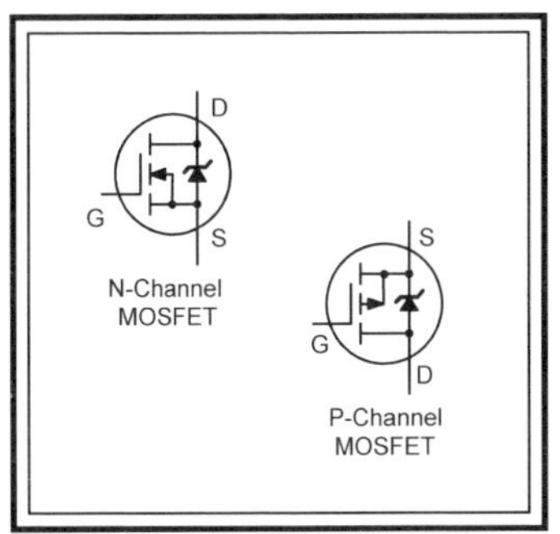


During "conduction", transistor devices exhibit what is called an "equivalent on-state resistance."

This means that when turned "on," the transistor does *not* act as a "perfect switch." There will be some power lost in the form of a voltage drop across the "switched" terminals (between the collector and emitter). In bipolar transistors (BJT's), this drop can be significant, especially in higher current applications.

MOSFET (Metal Oxide Semiconductor, Field Effect Transistor) technology has produced some *amazingly* low "on-state" resistances, resulting in extremely low voltage drops. They are significantly more efficient and closer to the "perfect" switch than most other types of semiconductors.

Depending upon the amount of current flowing through the device, typical on-state resistances of BJT's can be measured in the 100's or even 10's of ohms.

Many MOSFET's are capable of switching 60 amps of current (that's amps, not milliamps!) or more. At this high level of current flow, on-state resistances are typically measured in milliohms (thousandths of an ohm). This provides for some incredibly simple control circuit interfacing between low current devices (like a micon) and high current items like drive motors (perhaps on a mobile robot).

MOSFET's, like their BJT cousins, are also available in two primary types. These are known as "N-channel" and "P-channel" devices. And like the BJT, the P-channel types are generally more expensive and harder to find than the equivalent N-channel device.

The "Gate," "Drain," and "Source" are equivalent to the BJT's Base, Collector and Emitter, respectively. However, instead of causing a current to flow from the gate to the source, the switching action is caused by a mere "voltage presence" on the gate – there really is no direct *current flow,* which adds to their efficiency.

Figure 4.20 shows a typical driver circuit using an N-channel MOSFET. The lamp can be a high current type (several amps), but you may want to use separate power supplies because of the possibility of switching spikes.

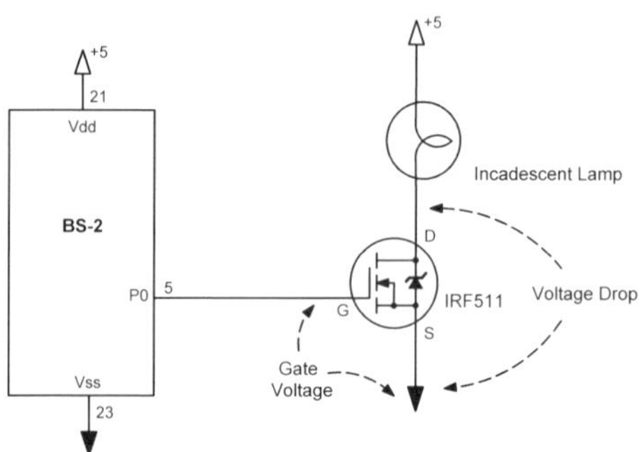


**Figure 4.20**

***A "high" signal applied to the gate results in a very low "on-state" resistance, which results in a low voltage drop***

**Code 4.20**

```
here:
low 0
pause 1000
high 0
pause 500
goto here
```

Figure 4.21 is a graph showing the allowable current through the drain of the MOSFET, versus the voltage applied at the gate. As you can see, since the "high" output of a microcontroller is typically less than 5 volts, the maximum amount of current that can flow through the device is never reached.

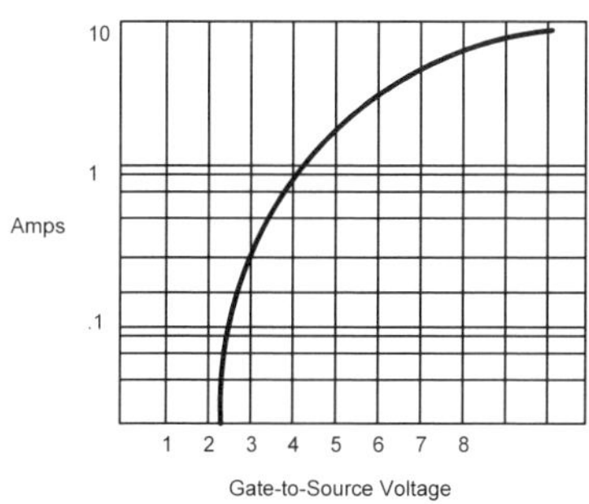


**Figure 4.21**
***The IRF511's transfer characteristics***

**Code 4.21**

```
"no code"
```

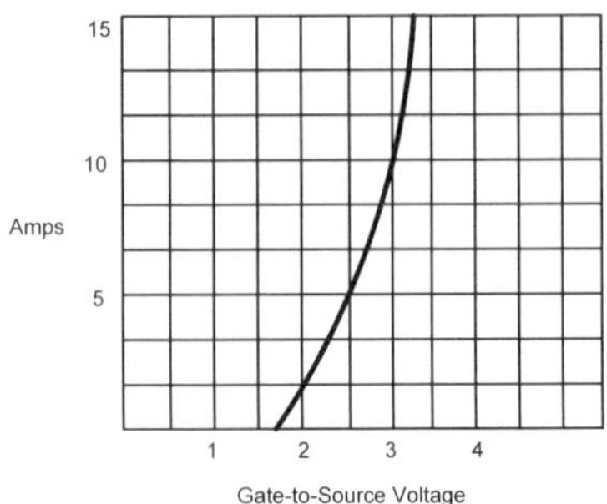


**Figure 4.22**
***A "logic-level" MOSFET, such as the RFP10P03L***

**Code 4.22**

"no code"

Figure 4.22 depicts a more recent development called a "logic-level" MOSFET. This device is capable of delivering full rated current with logic-level signals applied to its gate. Of course, you'll pay more for this capability and convenience.

As you can see in both graphs, the higher the voltage applied at the gate, the more efficient the "turn-on" of the MOSFET.

Since the IRF511 is not a "logic level" device, be sure to check its data sheet for allowable current versus gate voltage limits if you're planning on driving a high amperage device.

Figure 4.23 adds a simple "pull-up" resistor to the gate of the IRF511. This ensures that the voltage available to drive the gate is as high as the circuitry allows. With the Stamp,

this resistor is not really necessary because it (the Stamp) produces a logic "high" very close to its operating voltage (+5 volts).

If you use a buffer that doesn't have this high an output voltage (for example, the TTL series of IC's), then you should include the pull-up to ensure adequate "turn-on."

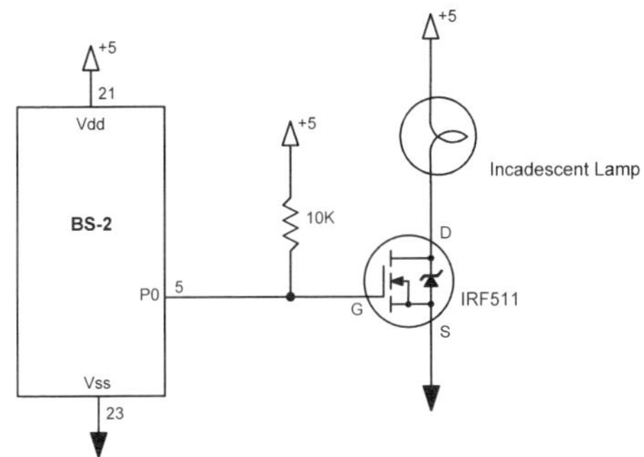


**Figure 4.23**

***Adding a pull-up resistor may be required if your circuitry doesn't provide a +5 volt "high" - in the case of the Stamp, the resistor is not required***

**Code 4.23**

```
here:
low 0
pause 100
high 0
pause 100
goto here
```

Since the (typical) maximum voltage that should be applied to a microcontroller is limited to no more than 5 volts, our turn-on efficiency is limited (when using a device like the IRF511). By using the circuit shown in Figure 4.24, you can "pull-up" the gate to a higher voltage, ensuring a more effective switching action.

The 7406 is a TTL type IC that is designed to operate on 5 volts. However, unlike most other TTL devices, the 7406 is capable of having its output lines connected to higher voltages. In the case of the 7406, this "higher voltage" is limited to no more than 30 VDC. The 7406 is called a "high voltage, open-collector inverting buffer." The 7407 is the non-inverting version of the same device.

By connecting the gate of the MOSFET to 12 volts (the same higher voltage that is driving the lamp) the transistor is capable of delivering its full rated power to the load. Be sure to use adequate heat sinking if you're running heavy loads in your project.

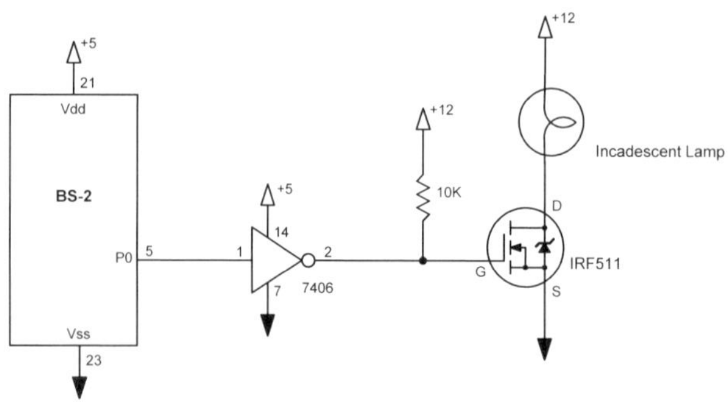


**Figure 4.24**
***A high-voltage applied to the gate using the 7406 IC***

**Code 4.24a**

```
low 0
stop
```

**Code 4.24b**

```
here:
low 0
pause 200
high 0
pause 200
goto here
```

P-channel MOSFET's are not as widely available as their N-channel counterparts, but can be quite useful in many different situations. For example, most automotive applications have a negative ground system. In this way, you can connect one side of an electrical device (such as a headlamp) to the frame of the car, and then "source" current through a single conductor wire.

Figure 4.25 shows a typical connection for a P-channel device that delivers a *sourced* voltage to the load, in this case a lamp. Note that driving the gate low turns on a P-channel MOSFET. Its operation is reverse of the N-channel devices.

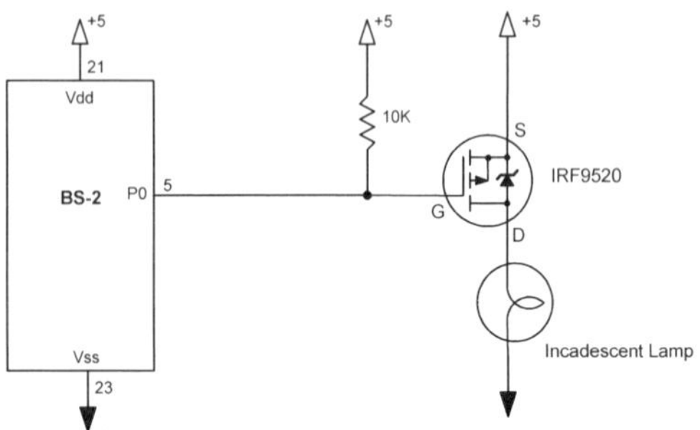


**Figure 4.25**
***Using a P-channel MOSFET to source voltage to a load***

**Code 4.25**

```
low 0        'turn on the lamp
stop
```

# Optoisolator Output

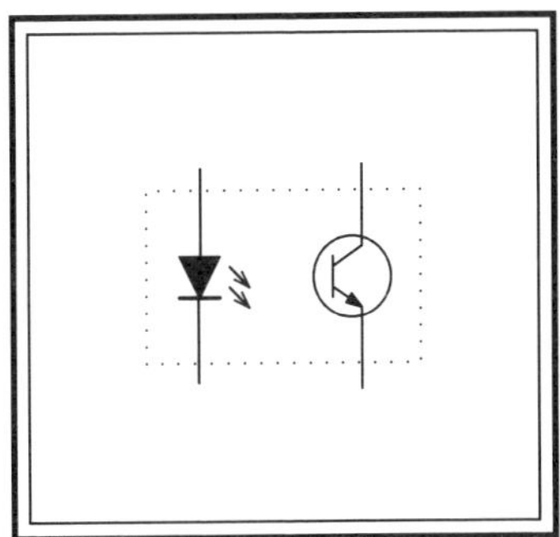

Optoisolators are essentially nothing more than an LED and a phototransistor contained within the same integrated package.

Some devices may have built-in resistors for current limiting through the LED. This feature, however, may limit you as to how much voltage you can use to drive the circuit.

Although Figure 4.26 uses the PS2501, you can substitute just about any other general-purpose device. Just be sure to not exceed the maximum ratings as outlined in its data sheet. By choosing the appropriate value of resistor you can apply virtually any voltage to the LED side of the device. Use ohm's law to solve for the appropriate value of the resistor.

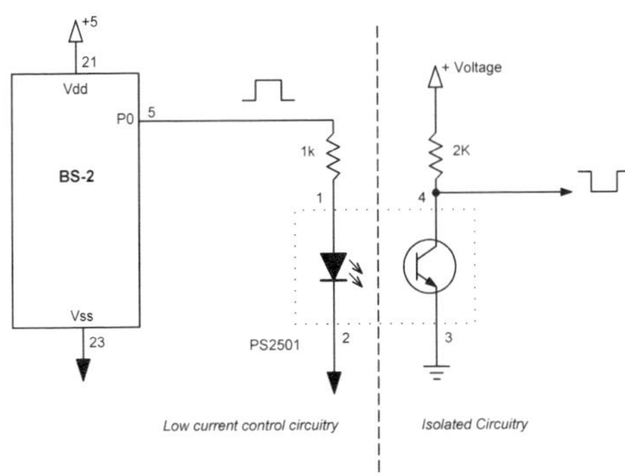


**Figure 4.26**
***A simple optoisolator circuit***

**Code 4.26**

```
here:
high 0
low 0
goto here
```

The circuit in Figure 4.27 triggers a 555 timer "one shot." Your program can cause an output pulse of a duration set by the pot and the 1.0 mf capacitor. In critical situations, your program merely starts the pulse, but doesn't have to worry about "returning later" to shut it off.

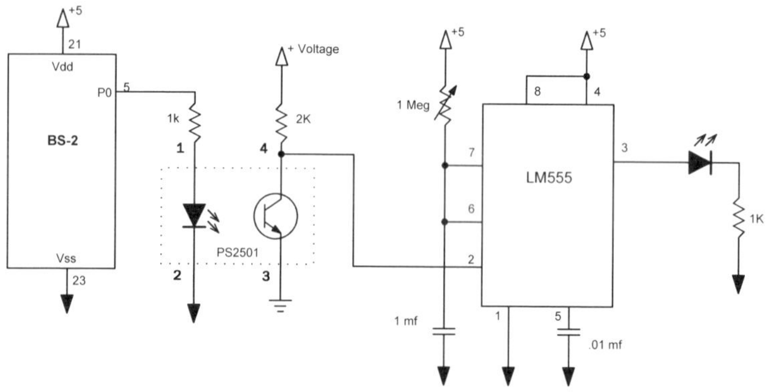


**Figure 4.27**
***A "trigger and forget" optoisolator circuit***

**Code 4.27**

```
low 0
high 0
low 0
here:
goto here
```

# Mechanical Relay

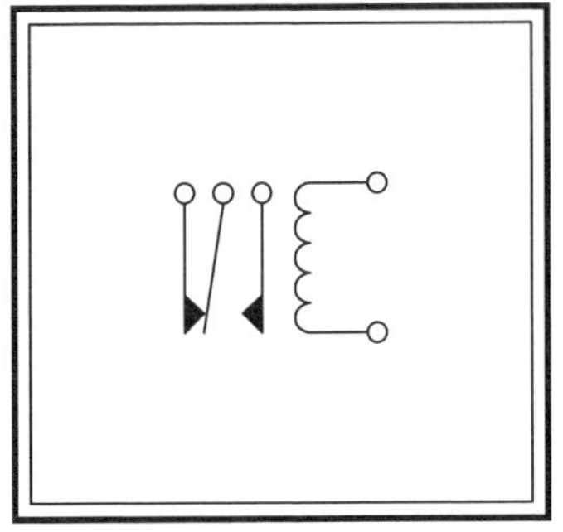

Mechanical relays come in a wide variety of shapes, sizes, configurations and ratings.

However, they all operate on the same basic principle, which is to operate a set of contacts by energizing an electro-magnetic coil. By passing an electric current through this coil, these mechanical contacts "open" and/or "close" associated circuits.

Although some mechanical relays may draw a low enough current to be directly operated by a microcontroller's I/O line, it's usually better practice to buffer the micon's control signal from the relay coil. When an electro-magnetic field collapses, a voltage spike is generated. This spike can damage sensitive MOS type circuits.

Figure 4.28 uses a 2N2222 transistor as the switching device. The "on" signal from P0 is "high," causing the transistor to conduct and sink the relay to ground. The magnetic field energizes, causing the relay contacts to actuate. Upon de-activation, the coil generates a reverse voltage spike. This spike is "snubbed" (or shorted out) by the 1N4001 diode. You may need to use a diode with a higher rating if your relay is a heavy-duty device.

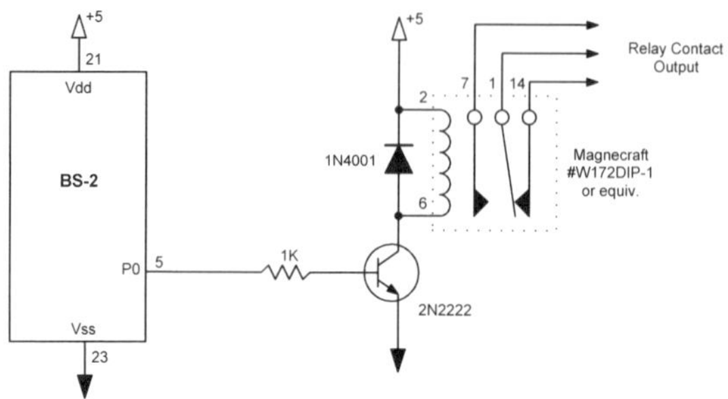


**Figure 4.28**
***Using an NPN transistor to drive a relay***

**Code 4.28**

```
here:
high 0
pause 1000
low 0
pause 2000
goto here
```

Some mechanical relays may have higher current demands than a small BJT can supply. Figure 4.29 will drive just about any DC relay you may require. The circuit uses a high voltage open collector 7407 buffer IC. The MOSFET is pulled high to 12 volts (depending on your choice of relay), ensuring maximum switching efficiency.

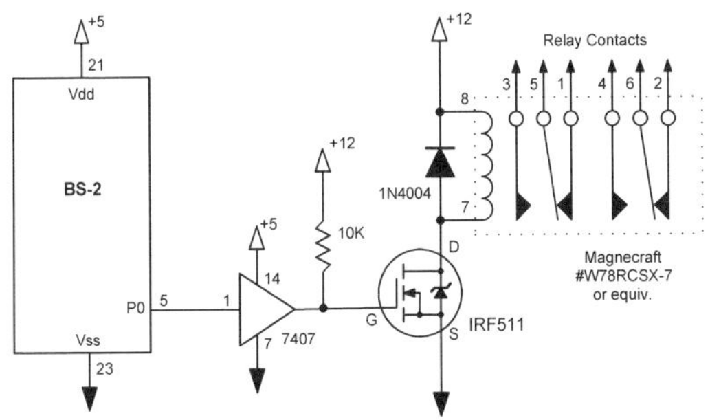


**Figure 4.29**

***A higher current relay driver circuit***

**Code 4.29**

```
here:
high 0
pause 1000
low 0
pause 2000
goto here
```

# Solid State Relay

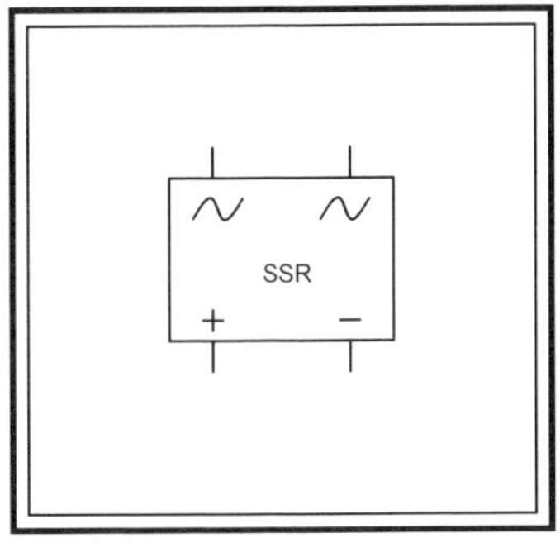


Solid State Relays (SSR's) are one of the simplest and safest methods to connect a microcontroller to a high voltage device.

They are available in many different configurations, but most operate with a built in optoisolation circuit. This is an advantage for two primary reasons.

First, the microcontroller and associated circuitry are electrically isolated from the high voltage device, providing a high degree of safety. Secondly, an optical isolation circuit requires a very low activation current – perfect for interfacing to a microcontroller.

Figure 4.30 is a "positive logic" control circuit. A "high" signal on P0 turns on the SSR.

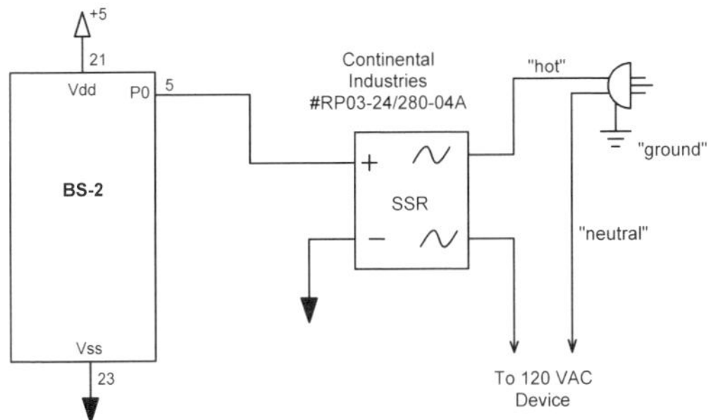


**Figure 4.30**

***For switching AC loads, it doesn't get much easier than this***

**Code 4.30**

```
high 0
stop
```

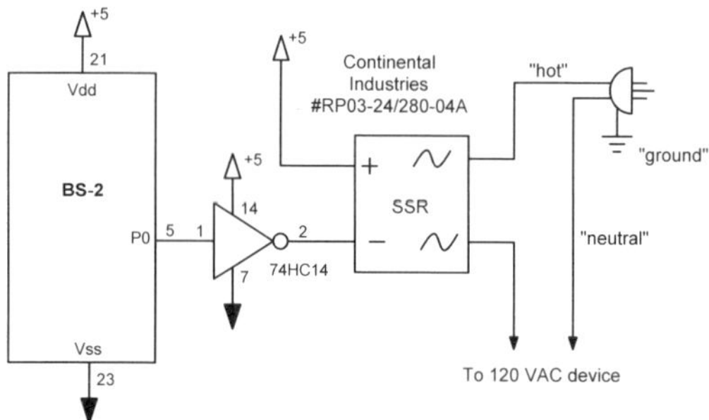


**Figure 4.31**
***The same "high" signal (which is inverted by the 74HC14) turns on the circuit, "sinking" the SSR to ground***

**Code 4.31**

```
high 0
stop
```

The circuit in Figure 4.31 also turns on the SSR with a high signal from the micon. But since the 74HC14 inverts the signal to a low, the relay is turned on by being "sunk" to ground.

Be sure to use adequate heat sinking techniques if using high load devices. As always, exercise extreme caution when working with high voltages.

# DC Motor

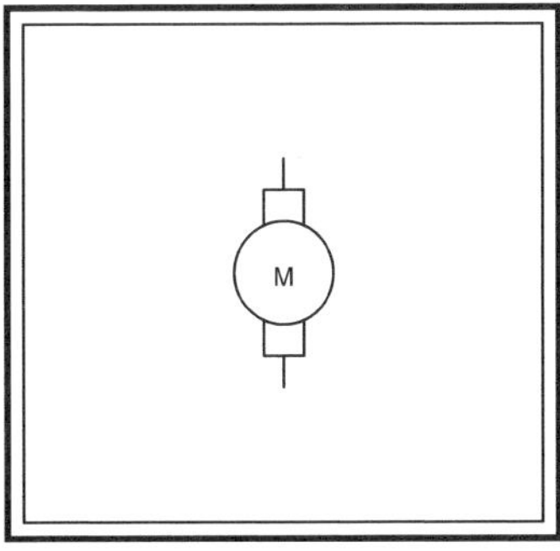


DC motors operate by allowing a current to flow through internal coils of wire. The subsequent interaction of the magnetic fields produces a rotary motion. This rotary action can then be used to manipulate things in the "real world."

DC motors, like relays and solenoids, will create a reverse voltage spike when the applied voltage is removed. For this reason, it is advisable to place a reverse biased diode across the motor terminals as shown in Figure 4.32. The diode shorts out the spike (generated by the collapsing magnetic field).

This circuit will control the on-off functions of a DC motor.

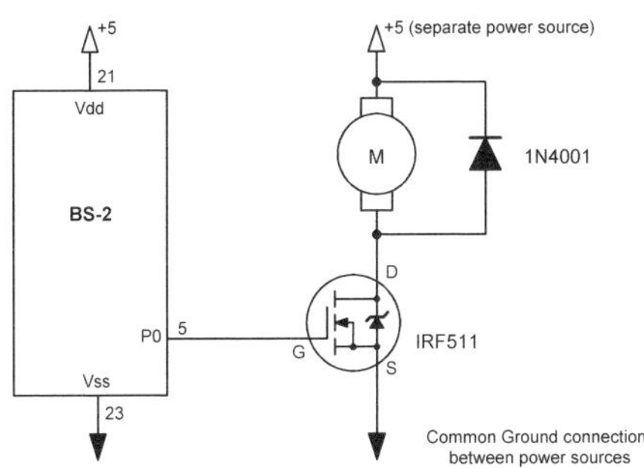


**Figure 4.32**
***A simple DC motor driver circuit***

**Code 4.32**

```
here:
high 0
pause 2000
low 0
pause 4000
goto here
```

The speed of the motor can be controlled by a technique called Pulse Width Modulation or PWM, for short. Figure 4.33 graphically depicts PWM. Use the circuit shown in Figure 4.32 and run Code 4.33 to see how the speed of the motor changes.

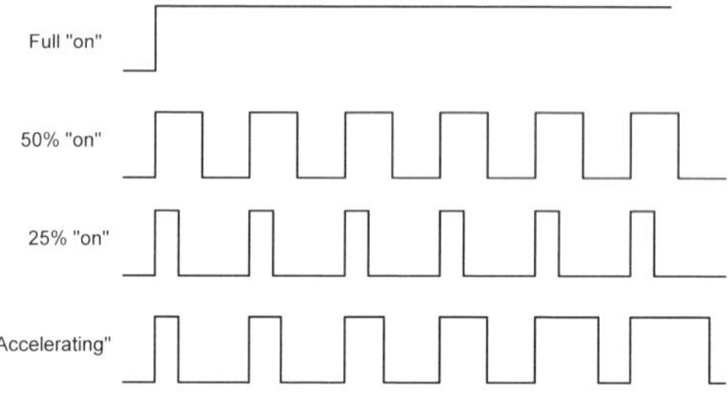


**Figure 4.33**
***Pulse-Width Modulation – "PWM"***

**Code 4.33**

```
x var word
y var word
here:
debug "This is full-on"
debug cr
high 0                    'ON all the time
```

```
pause 2000

debug "This is 50% on"
debug cr
for x= 1 to 200
high 0
pause 5                'ON for 5 milliseconds
low 0
pause 5                'OFF for 5 milliseconds
next

debug "This is 25% on"
debug cr
for x= 1 to 100
high 0
pause 5                'ON for 5 milliseconds
low 0
pause 15               'OFF for 15 milliseconds
next
pause 2000             'let the motor slow down

debug "This is accelerating"
debug cr
for y= 100 to 1
high 0
pause 15
low 0
pause y
next

goto here
```

A significant advantage in using the PWM technique is that *full* voltage is applied, if only for a short period of time. This means that even though the motor is not "on" all the time, maximum torque is still achieved.

By changing the width of the pulses, you can vary the speed of the motor. See Code 4.33 for a PWM "acceleration" example.

If you need to reverse the direction of a DC motor, one of the most common methods is shown in Figure 4.34. This is called an "H-bridge," and has been used extensively in bipolar transistor circuits.

Because of the wide availability and low cost of MOSFET devices, the efficiency of these types of circuits has increased dramatically. With on-state resistances measured in *milliohms*, MOSFET's lend themselves well to battery-operated circuits that require tight control of DC motors. Applications in the mobile robotics area are obvious.

Another advantage of an H-bridge circuit is that (especially with the low on-state resistance of MOSFET's) the circuit provides for a built-in "dynamic breaking" action. You can see in the schematic that if both N-channel devices are on, the motor terminals are effectively shorted to themselves. This may provide your project with a greater level of control.

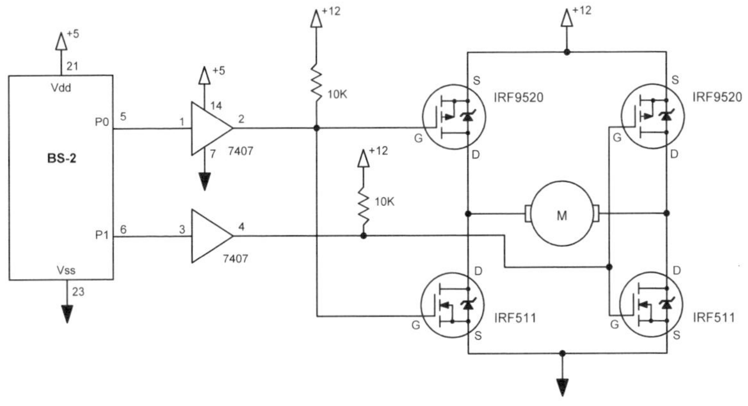


**Figure 4.34**
*A simple "H-bridge" DC motor driver circuit*

**Code 4.34**

```
here:

high 0
low 1
pause 2000
low 0
high 1
pause 2000

goto here
```

Figure 4.35 is the same basic circuit, but also includes optoisolation.

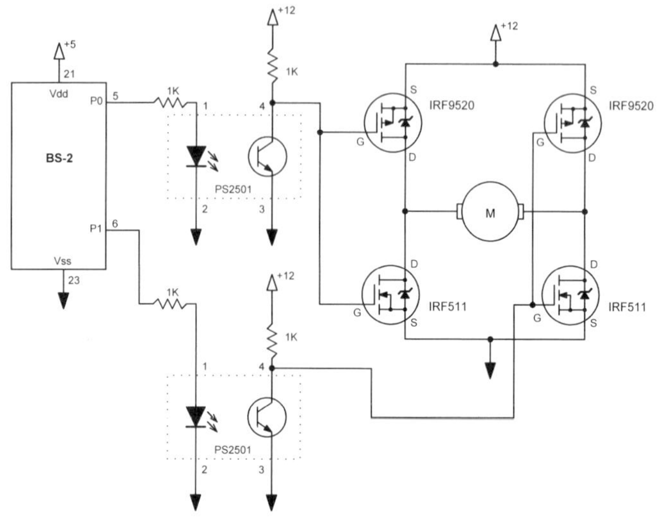


**Figure 4.35**

***Optoisolators provide glitch immunity from the high current motors. Be sure to use separate power supplies (and grounds) if required in your application***

**Code 4.35**

```
here:

high 0
low 1
pause 2000
low 0
high 1
pause 2000

goto here
```

# DC Servo

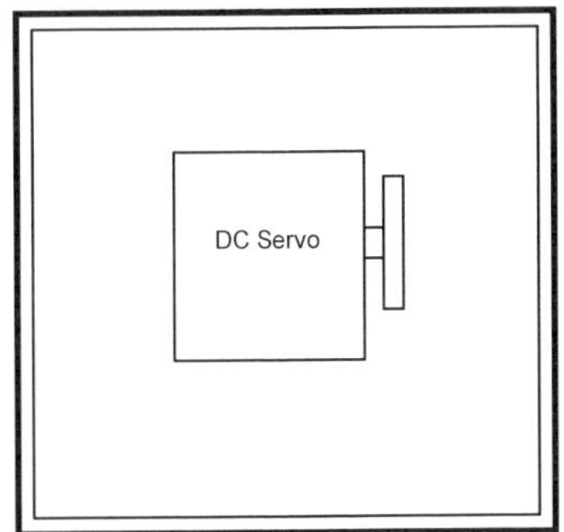


A DC servo is simply a motor with some additional control circuitry built into the motor housing itself.

Many of these devices are available for relatively low cost at hobby stores. These servos were originally designed for different types of radio-controlled devices such as model boats and airplanes. However, they are becoming increasingly popular in many different types of microcontrolled manipulative devices.

Servos are typically connected as shown in Figure 4.36. There are usually three connections to the device – *power, ground,* and *control.*

The voltage requirement is usually only 5 volts, but be sure that your power supply can deliver enough current. When a servo moves, there can be a large (over 1 amp) current requirement.

Depending on the capacity of your power supply, you may have to increase the size of the capacitor or use a separate power supply altogether. These power connections not only supply power to the motor itself, but also to the circuitry embedded inside, therefore clean power is a must.

"Control" is an input that expects a stream of pulses, as shown in Figure 4.37. Most servo control signals (for this "hobby" type of device) are designed to operate on pulse widths that vary between one and two milliseconds. Varying the pulse width between these two values causes the servo to move from one extreme to the other. This pulse needs to repeat about once every 10 milliseconds.

To hold its position, a servo must continue to receive the stream of pulses. If the steam is interrupted, there is no guarantee that the servo will remain in position - especially with any mechanical load. Without the pulse stream, the only thing that holds the servo in position is the internal friction of the motor itself. If there are any external forces on the servo when the pulse stream is interrupted, the servo may "lose" its place.

Since a microcontroller really isn't able to do more than one thing at a time, it would take some creative programming techniques to continue producing a pulse stream to hold that position, while simultaneously trying to accomplish other tasks. It's possible to be sure, but may make programming quite difficult.

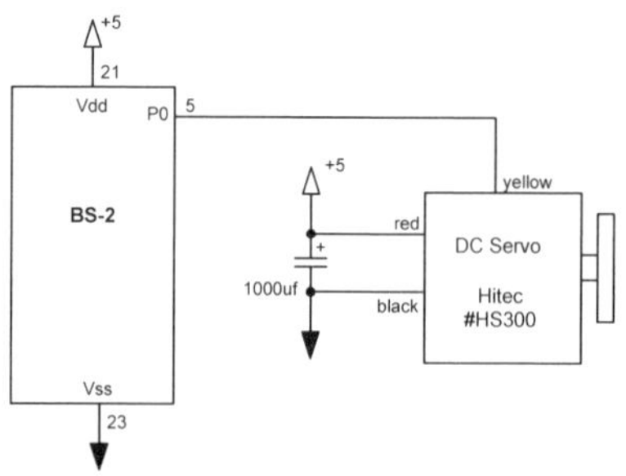


**Figure 4.36**
***A single I/O line is all that is required to drive a DC servo***

**Code 4.36**

```
x var word
y var word
output 0
here:

for y= 500 to 1000          '1 ms to 2 ms pulses
for x= 1 to 5               'send a string of (5) consistent
pulsout 0,y                 'pulses so the servo can keep
                            'up
pause 10
next
next
goto here
```

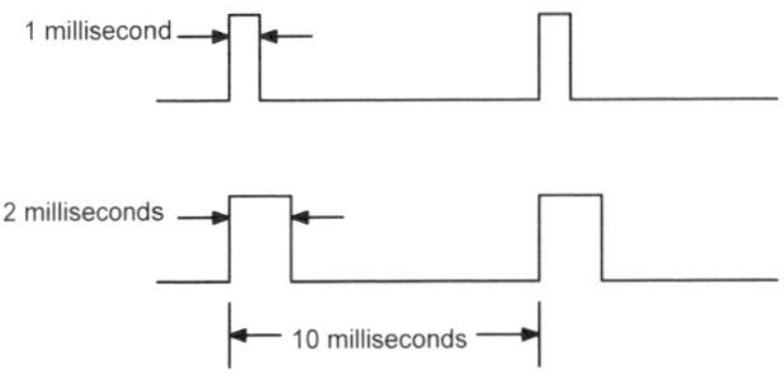


**Figure 4.37**
***Timing requirements of a typical DC servo***

**Code 4.37**

"no code"

Figure 4.38 solves this dilemma with hardware. This circuit is designed to produce a continuous stream of pulses that vary between one and two milliseconds.

The LM555 timer produces the pulses. However, instead of varying the pulse width with a conventional potentiometer, we're using a solid-state pot made by Dallas Semiconductor. The circuit uses three I/O lines from the micon to control the DS1804 pot, and one to control the 555 timer.

Although we're using a total of four I/O lines to control the servo (three more than Figure 4.36), we're eliminating the need for the program (embedded in the micon) to continually output the pulse stream.

The circuit adds hardware complexity but greatly simplifies programming, especially if the program needs to do other tasks while maintaining a servo's position.

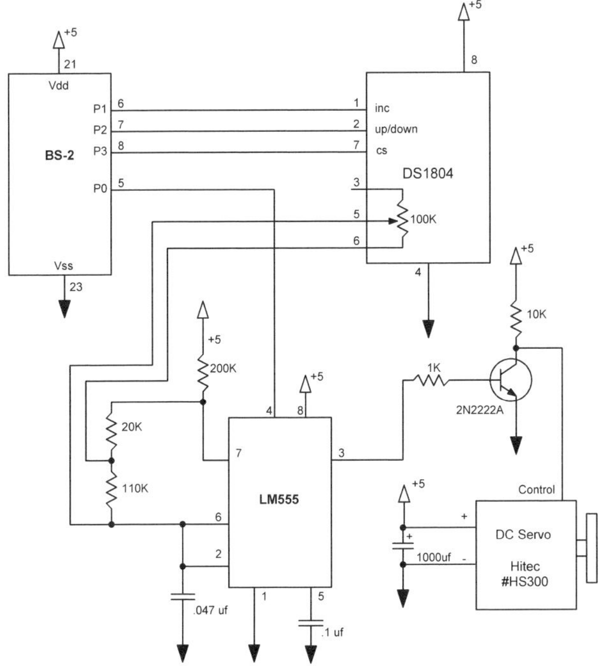


**Figure 4.38**
***Eliminating the need for a micon to produce the required pulse stream for a servo***

**Code 4.38**

```
x var word
y var word

here:
high 0                'enable 555

low 3                 'select DS1804
```

```
low 2                'set direction to counter-clockwise

for x=1 to 100       'reset the pot to "zero"
high 1
low 1
next

pause 1000           'give the servo time to respond

high 2               'set direction to clockwise
for y=1 to 90        'turn it in 100 discrete steps
pause 100
high 1
low 1
next

low 2                'set direction to counter-clockwise
for y=1 to 75        'rotate left
pause 100
high 1
low 1
next

debug "The micon is doing something else now, but the servo
remains steady"
debug cr
pause 3000
goto here            'back and forth
```

# AC Motor

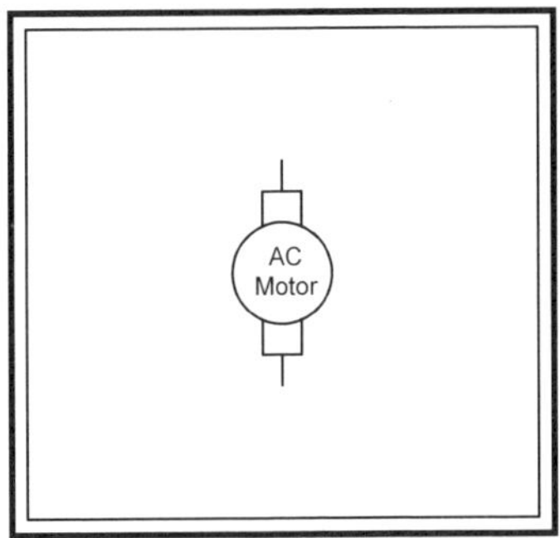


Most AC motors are designed to operate on 120 volts. Figure 4.39 depicts a typical motor connected to a solid-state relay.

You may want to add fuse protection to the "hot" leg for additional safety, especially if the source voltage is not fused or protected by an appropriate breaker. Always follow all applicable safety rules when working with "household" voltages.

Whenever you connect an AC load (motor, lamp, etc.) always be sure to switch the "hot" leg of power. "Hot" is designated by (usually) a black wire, and "neutral" is white.

Under normal circumstances, neutral has no potential in relation to ground (at the electric "service entrance", "ground" and "neutral" are connected together). Also, be sure to properly ground all equipment as specified by the motor specification data sheets. Be sure to check with a qualified electrician if you're unsure whether or not your facilities are wired correctly.

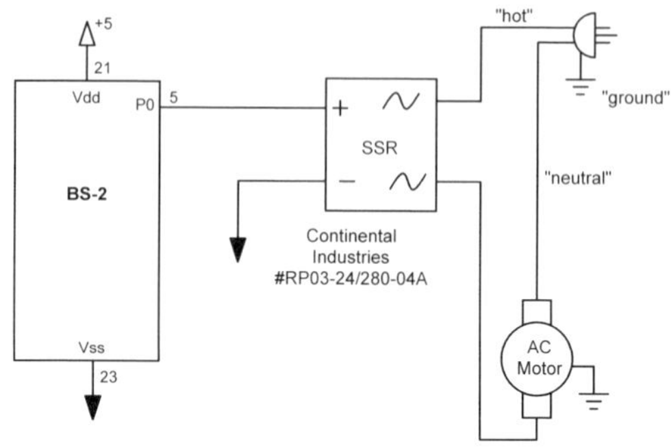


**Figure 4.39**
***An SSR circuit for turning on an AC motor***

**Code 4.39**

```
here:
high 0          'turn on the AC motor
pause 5000      'wait for 5 seconds
low 0           'turn off motor
pause 5000      'wait again
goto here       'do it again
```

Figure 4.40 shows how to use two SSR's to switch a 220-volt motor. In a "single-phase" 220 VAC system there are two "hot" wires (known as "legs.") Each leg must be switched and, because of the low current requirements of typical SSR's, both can be connected to the same I/O line of the microcontroller.

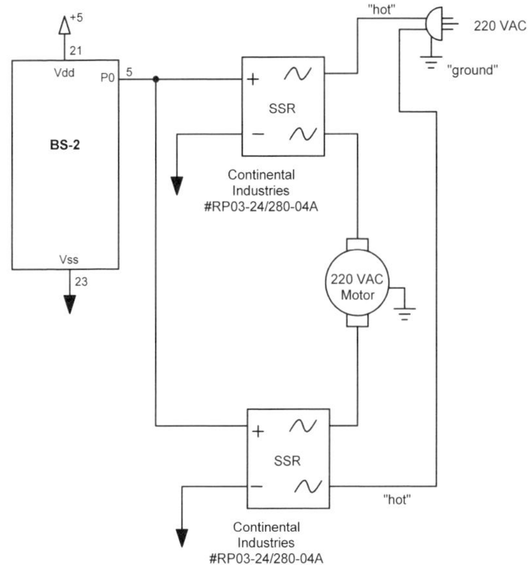


**Figure 4.40**
***Controlling a 220 VAC single-phase motor***

**Code 4.40**

```
here:
high 0          'turn on the 220 VAC motor
pause 5000      'wait for 5 seconds
low 0           'turn off motor
pause 5000      'wait again
goto here       'do it again
```

The National Electric Code is a comprehensive sourcebook for these types of applications. Another resource that provides a (more understandable) background on AC wiring and systems is “Wiring Simplified” by H.P. Richter and W.C.

Schwan (ISBN 0-9603294-4-7). This excellent book is also available at many home centers and hardware stores.

# Solenoid

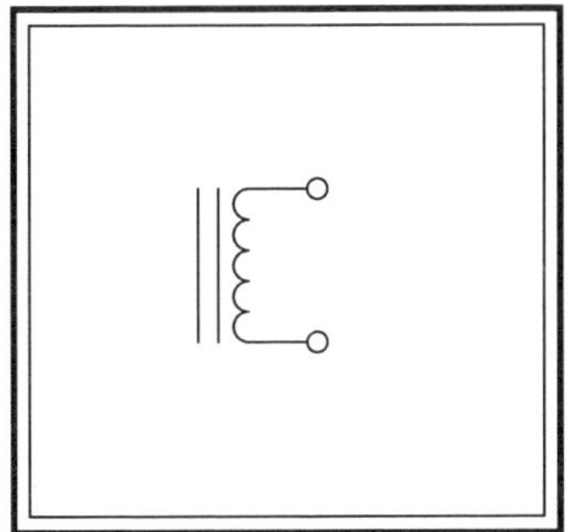

A DC solenoid operates in the same manner as a DC mechanical relay. They each use a wire coil to create a magnetic field, which causes a pivoting device (made of iron) to move when energized.

Figure 4.41 uses an NPN bipolar transistor to switch a low voltage solenoid. Be sure to include the reverse biased diode to snub the spike generated by the collapsing magnetic field.

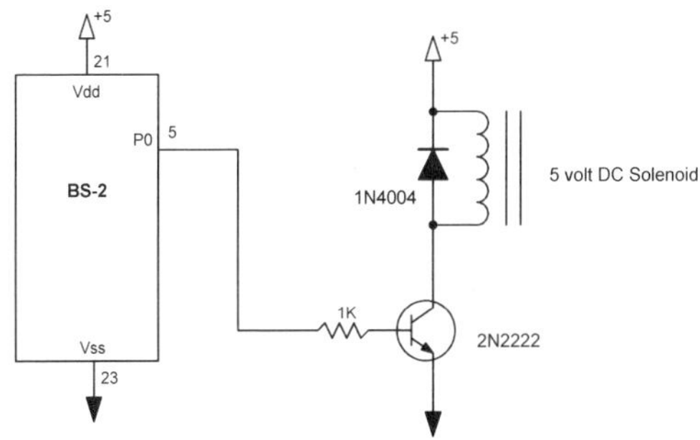


**Figure 4.41**
***A simple NPN transistor solenoid driver circuit***

**Code 4.41**

```
here:
high 0                    'turn on the solenoid
```

```
pause 1000          'wait for 1 second
low 0               'turn off solenoid
pause 1000          'wait again
goto here           'do it again
```

Figure 4.42 will drive a 12 to 24 volt solenoid. To ensure full "turn-on," it takes advantage of the 7406's high voltage output capability.

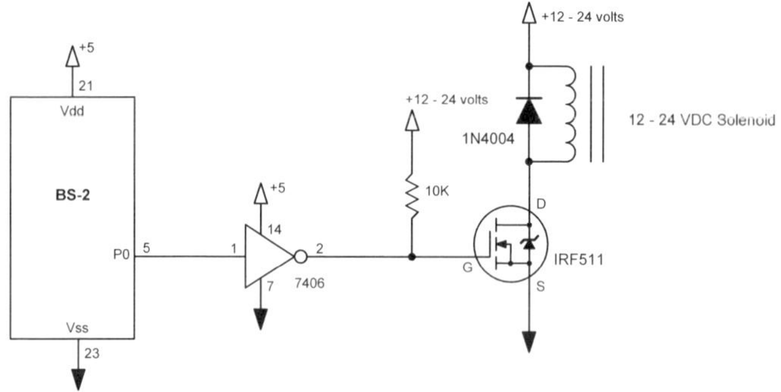


**Figure 4.42**
***Driving a high current solenoid***

**Code 4.42**

```
here:
low 0               'turn on the solenoid
pause 1000          'wait for 1 second
high 0              'turn off solenoid
pause 1000          'wait again
goto here           'do it again
```

When you really need a lot of linear "pull power," you can use a 120 VAC solenoid, as shown in Figure 4.43. As always, exercise caution when working with higher voltages.

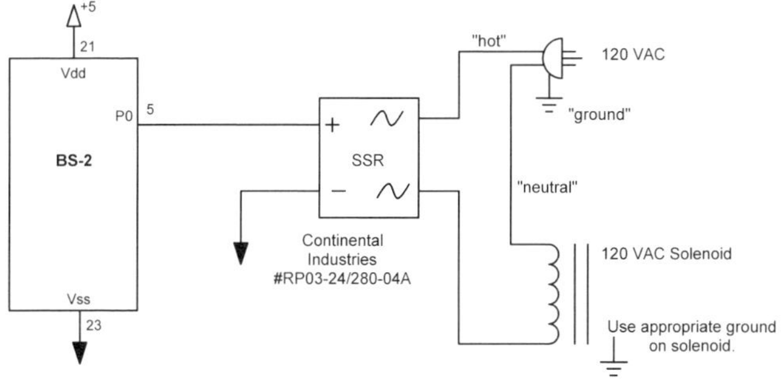


**Figure 4.43**
***Driving a 120VAC solenoid***

**Code 4.43**

```
here:
high 0          'turn on the solenoid
pause 2000      'wait for 2 seconds
low 0           'turn off solenoid
pause 2000      'wait again
goto here       'do it again
```

Many outdoor sprinkler systems operate on 24 VAC. They can be driven with this circuit as well (most AC SSR's operate on voltages ranging from 24 to 120 or 240 VAC.) Be sure to follow the data sheet for the device you're using.

# Speaker

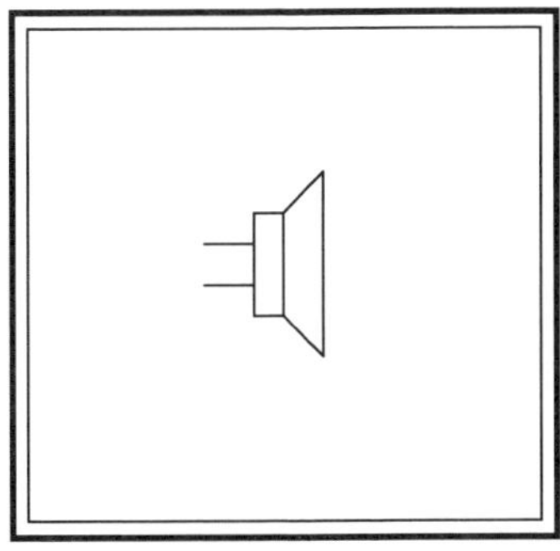

Humans can normally hear frequencies from about 15 hertz to 15,000 hertz. This is known as the "audible spectrum."

These "oscillations" are measured in "cycles per second" - more commonly known as "hertz."

A microcontroller such as the Stamp may operate internally at very high frequencies – well beyond our ability to "hear" them. For example, the internal clock rate of most microcontrollers is over 1,000,000 hertz (one megahertz).

Because the rate at which the I/O lines can "toggle" is completely under our program control, it stands to reason that we should be able to generate frequencies that we can hear.

Since most commonly available speakers have a relatively low resistance value (typically 8 ohms), it's prudent to use a current-limiting resistor if you're connecting it directly to an I/O line, as shown in Figure 4.44.

Code 4.44a generates a continuous stream of pulses causing the diaphragm to move air, resulting in sound.

The Stamp has some other unique commands that can generate different types of sounds as demonstrated by Code 4.45b. Check out the BASIC Stamp Manual for more information on these features.

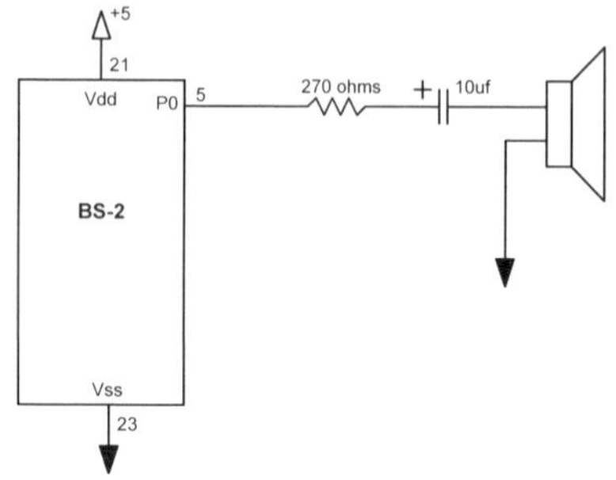


**Figure 4.44**
***A simple speaker connection***

**Code 4.44a**

```
here:
high 0
low 0
goto here
```

**Code 4.44b**

```
x var word
here:
for x=50 to 1
high 0
low 0
pause x
next
goto here
```

Figure 4.45 adds a MOSFET for greater volume.

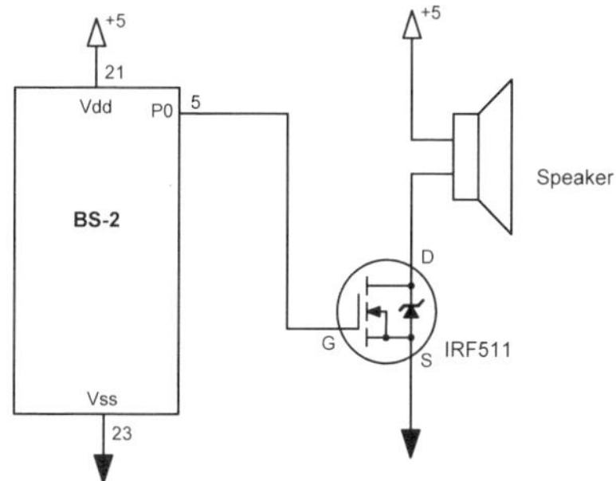


**Figure 4.45**
***A MOSFET can increase volume***

**Code 4.45a**

```
x var byte
here:
for x = 1 to 100
high 0
low 0
next
pause 500
goto here
```

**Code 4.45b**

```
freqout 0,1500,500,1000
pause 500
dtmfout 0,100,50,[5,5,5,1,2,1,2]
pause 500
freqout 0,3000,100,500
stop
```

Most microcontrollers are really only capable of doing one thing at a time. To create sound through the speaker, a microcontroller must output a continuous stream of pulses.

This means that the micon may not be able to "create sound" and, at the same time, monitor inputs or control other outputs.

Figure 4.46 uses a 555 timer (connected as an oscillator) to create a pulse stream. Pin 4 of the 555 is "reset" control. When pulled high, the 555 sends a stream of pulses to the speaker. P0 is used solely as an enable line, thereby allowing the microcontroller to accomplish other tasks.

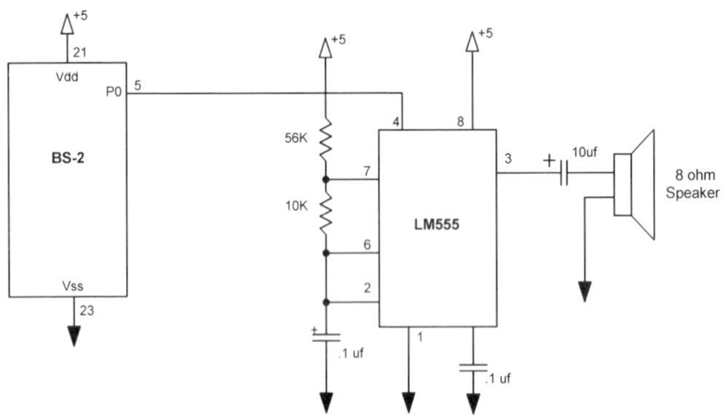


**Figure 4.46**
***Rather than the sound being generated by the micon, you can have an oscillator do the work instead of the program***

**Code 4.46a**

```
high 0
stop
```

**Code 4.46b**

```
here:
high 0
pause 500
low 0
pause 500
goto here
```

If you combine Figure 4.46 with a digital potentiometer (Figure 4.47) you can change the frequency of the 555 oscillator. Simply substitute the digital pot output pins (5 & 6) in place of the 56K resistor in Figure 4.46.

# Digital Potentiometer

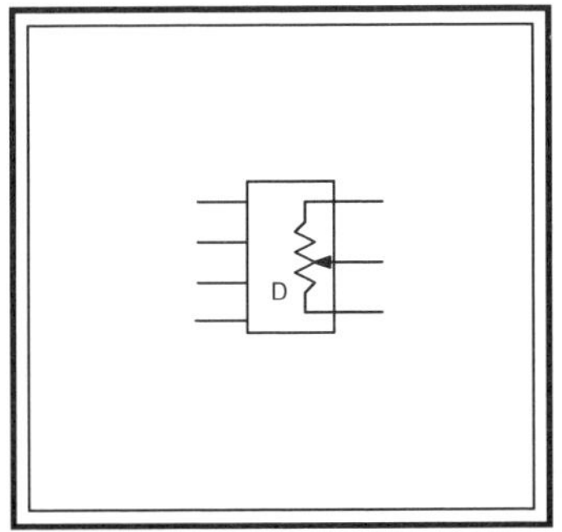


From a microcontroller's point of view, potentiometers are generally considered input devices. In this case however, we're going to have the micon control the pot.

A digital pot is designed to provide the same electrical characteristics as a "standard" pot but, instead of a mechanical action, it relies on digital signals from a microcontroller to change the resistance value.

Figure 4.47 shows the basic connections between the Stamp and the DS1804-100. This device is available in several different values and resolutions. Check the manufacturer's data sheet for more details.

The DS1804 has 100 discrete "positions" for the wiper. Other devices are available with different resolutions (more or less number of discrete steps on the wiper).

The basic device has three inputs: "increment," "up/down," and "chip select."

Increment is essentially a clock input. Each pulse applied to this pin will cause the wiper to move in relation to the resistive element (contained within the IC itself). Up/down determines which direction the wiper will move, and chip select enables the device to respond to the other control signals.

Connect an ohmmeter between pins 3 and 5 on the DS1804, as you run Code 4.47.

A program embedded in the microcontroller can now operate any circuit that you may have previously controlled with a “manual” pot.

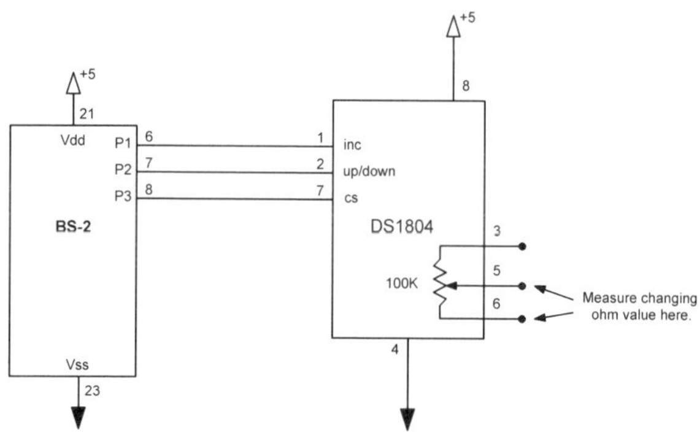


**Figure 4.47**
***The Dallas Semicondutor DS1804 digital potentiometer***

**Code 4.47**

```
x var byte

low 3                   'select DS1804
low 2                   'set direction to counter-clockwise

for x=1 to 100          'reset the pot to "zero"
high 1
low 1
next

here:
high 2                  'set direction to clockwise
for x=1 to 100          'step it 100 discrete times
pause 1000              'view it on the ohm meter
```

```
high 1
low 1
next

low 2                    'set direction to counter-clockwise
for x=1 to 100           'step it 100 discrete times
pause 1000
high 1
low 1
next
goto here
```

# Operational Amplifier

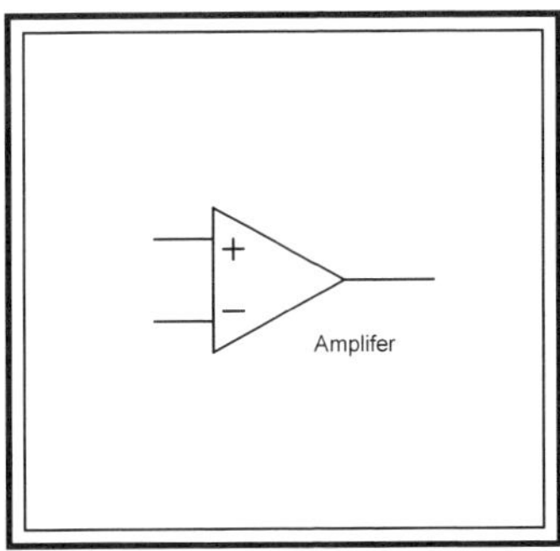


Operational amplifiers can be set to different gain levels by adjusting their associated resistor combinations. The circuit shown in Figure 4.48 is just a sample of the many potential applications you can develop using a digital pot as the "gain" adjustment.

Under program control, your microcontroller can adjust the gain of the circuit as required by the application.

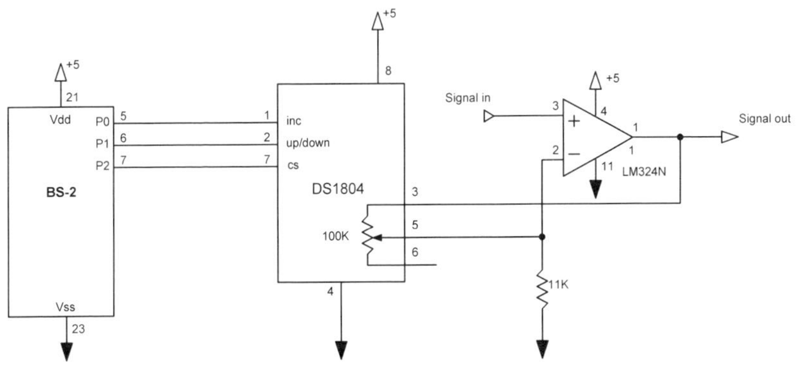


**Figure 4.48**
***Digital control of an analog op-amp***

**Code 4.48a**

```
x var byte
low 2                    'select DS1804
```

```
high 1                  'set direction to clockwise

for x=1 to 100          'reset the pot to "zero"
high 0
low 0
next

low 1                   'set direction to counter-clockwise
for x=1 to 8            'set gain to approximately '2'
                        'view it on the o'scope
high 0
low 0
next
stop
```

**Code 4.48b**

```
x var byte
here:
low 2                   'select the DS1804
high 1                  'set direction of the wiper movement
for x = 1 to 100        'move the wiper all the way to an end
high 0
low 0
next

low 1                   'change direction of the wiper movement
for x = 1 to 16         'ramp the gain of the op-amp
high 0
low 0
next
goto here
```

Try connecting a 1 volt signal to the input of the LM324, and look at the output of pin 1 with an oscilloscope as you run Code 4.48b. You'll see stepped amplification levels as the program changes the pot's wiper position.

Chapter 5

# Other Circuits

# Getting More Input

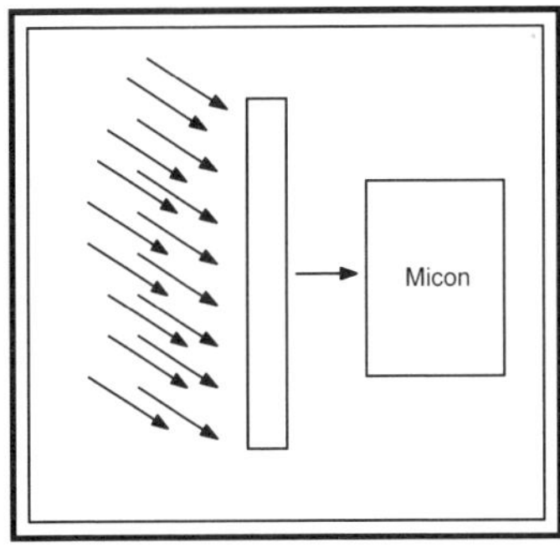


Sooner or later your project will require more I/O lines than your microcontroller has available. You are then faced with either adding another micon to your project, or creating some additional circuitry that "expands" the number of I/O lines.

In some cases it may be beneficial to add another controller to the system due to software and programming constraints.

However, if programming capacity is not an issue, it's usually less expensive to add (and easier to integrate) a circuit such as that shown in Figure 5.1. This circuit provides 8 separate inputs, but only uses 4 of the microcontroller's I/O lines.

The 74HC151 has 8 digital inputs (D0-D7), any one of which can be selected to "flow through" the device to the "Y" output. The three-bit binary value presented on inputs "A," "B," and "C" determines which input is passed through. "A" is the least significant bit, therefore a value of "001" will result in "D1" being selected.

Each of the inputs can be used just like a normal "input" line on a microcontroller. A simple switch is shown on D0 in the schematic, but any other type of sensor input can be attached. In fact, you could connect a myriad of different devices – a different type to each input - depending on your application. Unused inputs should be connected to ground.

Note that all inputs on the 74HC151 are actually channeled through the "Y" output and are "read" on P3 (on the Stamp). Your program must keep track of which three-bit address (on P0-P2) is responsible for the delivered input. Also, the selected input signal is inverted as it makes its way through the '151. Be sure your program accounts for this. If you don't want an inverted signal, use the output on pin 6 (on the 74HC151.)

Code 5.1a reads the value of input "D0." Code 5.1b reads the value of "D1." The address (created by the binary value on P0-P2) was changed from "000" to "001."

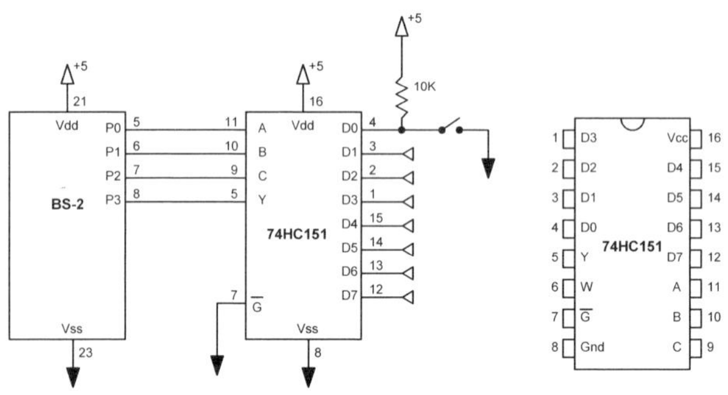


**Figure 5.1**
***Getting 8 inputs using only 4 I/O lines***

**Code 5.1a**

```
x var bit

low 0          'A=0, this 3-bit address looks at D0
low 1          'B=0
low 2          'C=0
here:
x=in3
```

```
debug ? x
goto here
```

**Code 5.1b**

```
x var bit

high 0          'A=1, this 3-bit address looks at D1
low 1           'B=0
low 2           'C=0

here:
x=in3
debug ? x
goto here
```

By adding two additional control lines, we can expand up to 16 input lines, as shown in Figure 5.2.

This circuit utilizes the “enable” feature of the 74HC151. Toggling P4 selects which ‘151 is to be used, and the 3 bit address present on A, B and C determines which one of the 8 channels is selected. Outputs from both ‘151's are connected to an OR gate, which allows either chip to deliver data to the microcontroller.

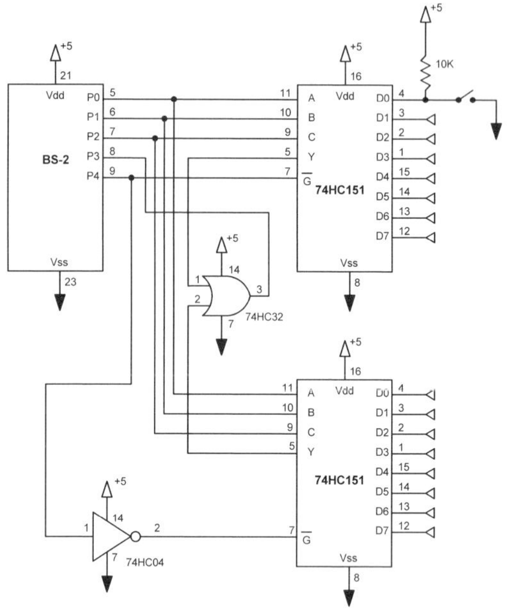


**Figure 5.2**
***16 inputs using 5 I/O lines***

**Code 5.2a**

```
x var bit

low 0          'this 3-bit address selects which input
low 1          '000 = D0 input
low 2

low 4          'this selects the first '151

here:
x=in3          'reads the value of the input
debug ? x
goto here
```

**Code 5.2b**

```
x var bit

low 0                   'this 3-bit address selects which input
high 1                  '010 = D2 input
low 2

high 4                  'this selects the second '151

here:
x=in3                   'this reads the value of the input
debug ? x
goto here
```

Figure 5.3 takes this same concept to the next level by using four separate 74HC151's. Remember to ground all unused inputs (on the '151's) to prevent unwanted glitches in the system.

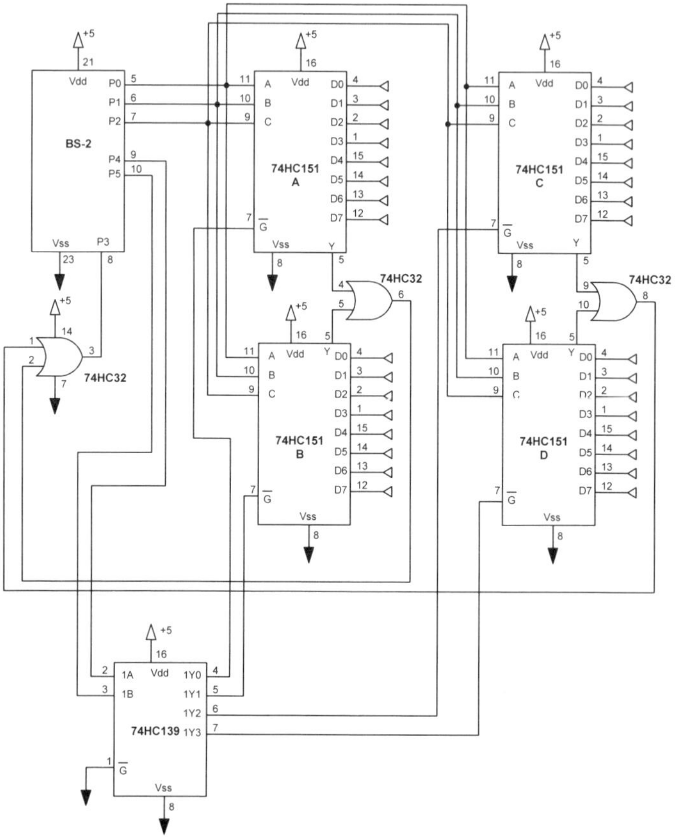

**Figure 5.3**
***32 inputs using only 6 I/O lines***

**Code 5.3a**

```
x var bit
low 0                'this 3-bit address selects which input
low 1                '000 selects D0
low 2
```

```
low 4                    'this selects the first '151
low 5                    'this is binary '00'

here:
x=in3                    'this reads the value of the input
debug ? x
goto here
```

**Code 5.3b**

```
x var bit

high 0                   'this 3-bit address selects which input
low 1                    '001 selects D1
low 2

high 4                   'this selects the second '151
low 5                    'this is binary '01'

here:
x=in3                    'this reads the value of the input
debug ? x
goto here
```

**Code 5.3c**

```
x var bit

low 0                    'this 3-bit address selects which input
high 1                   '010 selects D2
low 2
low 4                    'this selects the third '151
high 5                   'this is binary '10'

here:
x=in3                    'this reads the value of the input
debug ? x
goto here
```

**Code 5.3d**

```
x var bit

low 0           'this 3-bit address selects which input
high 1          '110 selects D6
high 2

high 4          'this selects the fourth '151,
high 5          'this is binary '11'

here:
x=in3           'this reads the value of the input
debug ? x
goto here
```

The three-bit address operates as before, but now we need to enable only one of the four chips at a time. To accomplish this we've added a "2 line to 4 line decoder" (the 74HC139). This device outputs a low on one of four outputs. These signals are then used as "chip enables" for each 74HC151.

This circuit may be useful if your project requires the intermittent monitoring of many different inputs, such as that which may be found on a burglar alarm system, etc.

# Expanding Output

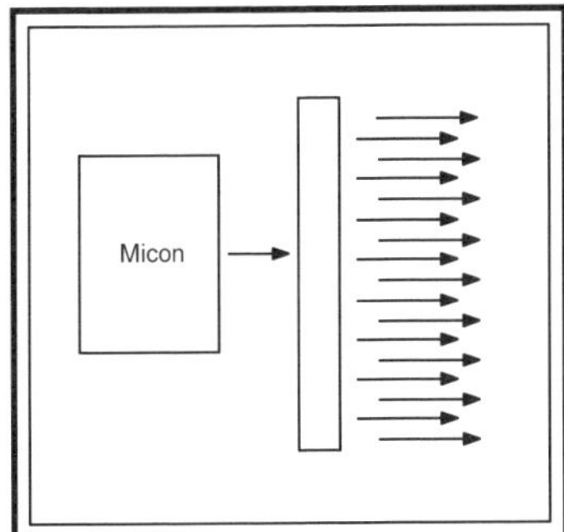


Adding digital outputs can be accomplished in many ways. If your project only needs to control a single line at a time, then you could use the circuit shown in Figure 5.4.

The 74HC154 is a "4 line to 16 line decoder" IC. This simply means that the device will take a 4 bit binary value and actuate the appropriate output (0-15). The disadvantage, of course, is that only one of the 16 outputs can be "low" at any one time. This circuit is well suited for applications such as "running light" or sequencing devices.

Only three LED's are shown in the schematic, but you get the idea. Any device that can be operated with a "low" signal can be connected to each of the 16 outputs. Of course, you can add inverters if required.

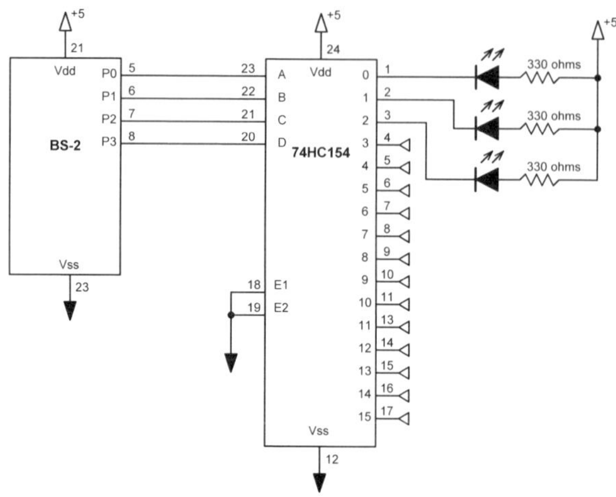


**Figure 5.4**
***A 16 output sequencer / selector***

**Code 5.4**

```
x var byte

here:
low 4               'this enables the '154
dira=15             'set up 4 bits as outputs
for x=0 to 15       'count to 15 in binary
outa=x              'put the value out to the '154
pause 100           'slow down so we can see it

next
goto here
```

Figure 5.5 uses a 74HC138 “3 line to 8 line decoder.” Its operation is very similar to the 74HC154.

The outputs are addressed by the 3-bit code as presented on “A,” “B,” and “C” (with “A” being the least significant bit). “E1” must be high for the ‘138 to operate. See Code 5.5a.

You can also use this circuit as a “serial data output selector,” as demonstrated by Code 5.5b.

For example, by setting the address to “010,” output 2 is driven low. Now as shown in the code, each time “E1” is toggled, the level on “output 2” follows (as inverted data). Thus, by setting the appropriate address, you can send serial data *out* through any selected output line.

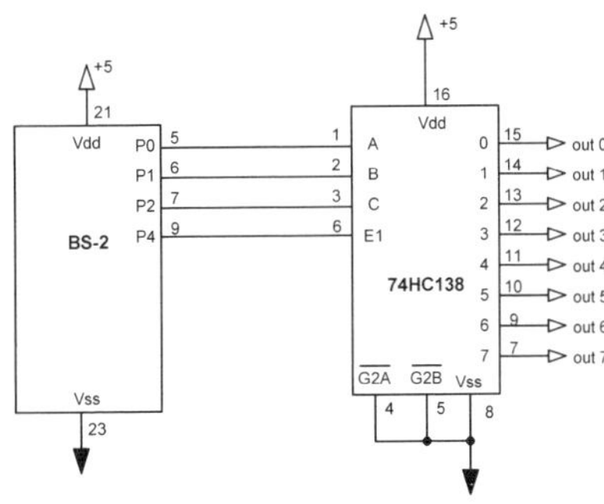


**Figure 5.5**
***8 outputs from 4 control lines***

**Code 5.5a**

```
x var byte

here:
high 4                  'this enables the '138
dira=7
for x=0 to 7
outa=x
```

```
pause 100
next
goto here
```

**Code 5.5b**

```
high 4              'this enables the '138
dira=7
outa=2              'data will go out on "out 2"

here:
toggle 4            'this can be your serial data output
pause 100
goto here
```

The circuitry can be expanded to that shown in Figure 5.6. This schematic provides 24 outputs using only five I/O lines on the microcontroller.

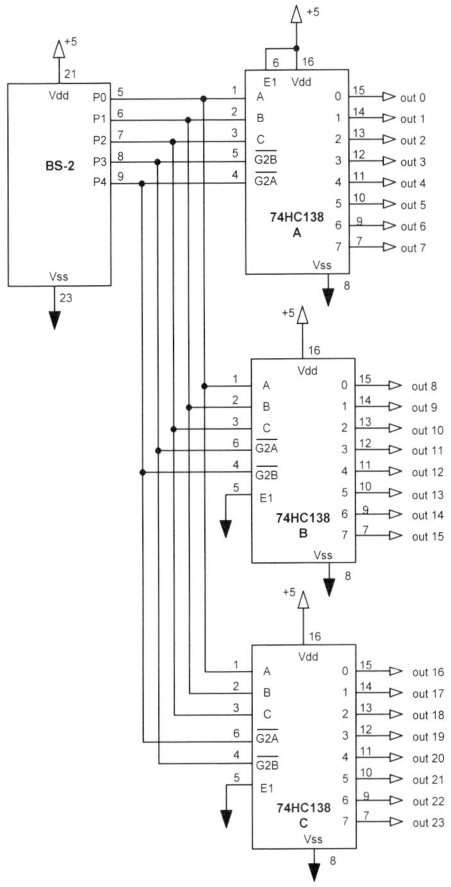


**Figure 5.6**
***24 outputs using only 5 I/O lines***

**Code 5.6**

```
x var byte
dirl=255                'set up lower byte as outputs
here:
```

```
for x=0 to 23          'sequence through 24 outputs
outl=x
pause 100
next

goto here
```

If your application requires the on/off control of individual bits, then you can use the circuit shown in Figure 5.7.

The 74HC164 is an 8 bit "shift register." This device receives serial data from a single I/O line and outputs the corresponding bits in parallel. Unlike the prior circuits, this circuit allows any combination of bits to be presented as outputs simultaneously.

Since this device has no provision for turning "off" its output lines, care should be exercised when using this circuit. As you "clock" the serial data into the '164, each output will toggle in sequence as the data "flows" through the register.

This can be readily demonstrated by increasing the Pause command to something "visible" (slow enough to watch the LED's respond) in Code 5.7. As each bit is clocked into the device, the outputs change accordingly.

In many non-critical applications this may not be a problem. However, if you were to connect a Solid State Relay to one of these pins, and then change one of the other bits to a different value, for a split-second the SSR may switch on (or off) as new data ripples through the register. Your circuit could actuate something at the wrong time. Be careful!

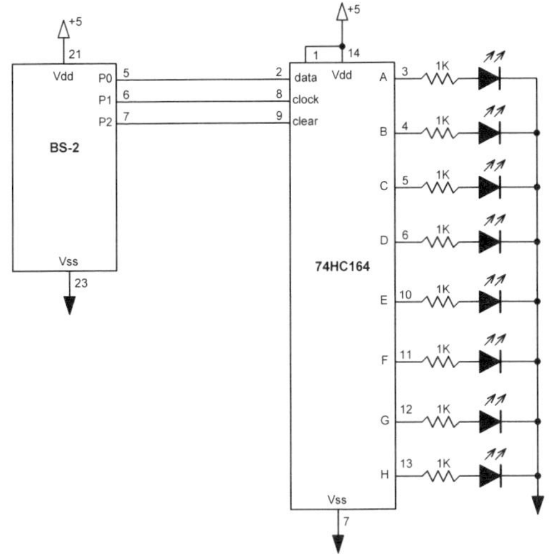


**Figure 5.7**
***An 8-bit parallel out serial shift register***

**Code 5.7**

```
x var byte
low 0                  'set data line to 0

here:
low 2                  'clear the '164 to 0000
high 2                 'enables the '164 to receive a new value

for x=1 to 8           'send 8 bits through a single I/O pin
toggle 0               'change the bit - this is the data bit
high 1                 'clock this bit into the '164
low 1
pause 100
next                   'clock in a total of 8 bits
pause 3000             'wait here with this 8 bit value
goto here              'go back and do it again
```

Since the shift register changes its output bits *as* the value is being clocked through the device, your circuitry may attempt to respond to these quick transitions. If this is not acceptable, you can add a flip-flop, as shown in Figure 5.8.

This circuit loads the data into the *input* side of the 74HC374 D-type flip-flop. After the data has "shifted out" of the '164 and into the '374, the P3 I/O line strobes the data through to the '374 outputs. This 8 bit transfer happens in parallel, thereby eliminating any "ripple" problems.

A new 8 bit pattern can then be loaded into the '374, without disrupting the present state of the outputs.

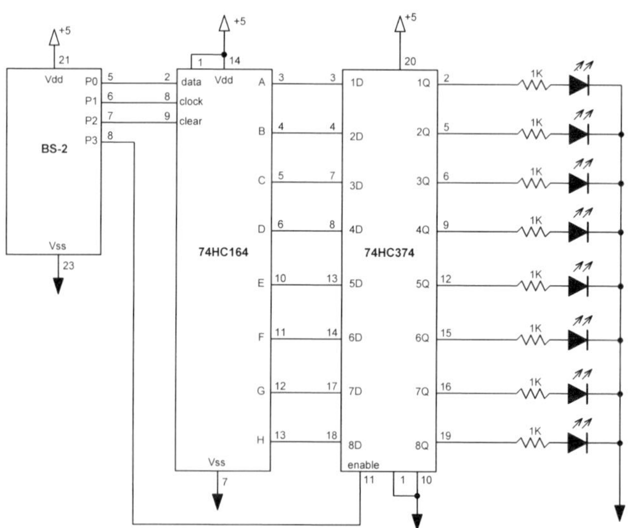


**Figure 5.8**
***The addition of a "flip-flop" allows data to be sent through simultaneously***

**Code 5.8**

```
x var byte
low 0                   'set data line to 0

here:
low 2                   'clear the '164 to 0000
high 2                  'enables the '164 to receive a new value
high 3                  'latch 0000 into the '374
low 3                   'disable the latch again

debug "I just cleared the register and the latch to 0000"
debug cr
pause 2000

for x=1 to 8            'send 8 bits through a single I/O pin
toggle 0                'change the bit - this is the data bit
high 1                  'clock this bit into the '164
low 1
next                    'clock in a total of 8 bits

debug "I just loaded 01010101 into the shift register"
debug cr
pause 2000
high 3                  'latches 8 bits from the '164 into '374

debug "I just latched the data through to the 374 outputs"
debug cr
pause 2000

low 3                   'disable the latch again
debug cls
pause 1000
goto here
```

Figure 5.9 takes this concept to the next level by adding a second 74HC374. Everything else in the circuit operates as before, but now your program can select which one of the '374's will "accept" the data. See Code 5.9.

Using two additional I/O lines results in 32 outputs, as shown in Figure 5.10. Each of these outputs are individually addressable, and we've only used seven I/O lines on the micon.

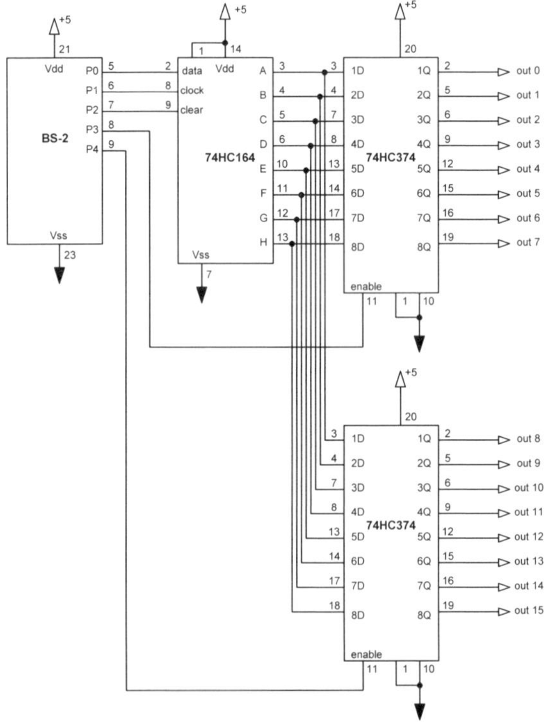


**Figure 5.9**
***16 output lines from only 5 I/O lines***

**Code 5.9**

```
x var byte
low 0                   'set data line to 0
low 3                   ‘disable the first ‘374
low 4                   ‘disable the second ‘374
here:

'This is for the first '374 latch
low 2                   'clear the '164 to 0000
high 2                  'enables the '164 to receive a new value
high 3                  'latch 0000 into the '374
low 3                   'disable the ‘374 again

debug "I just cleared the register and the latch to 0000"
debug cr
pause 2000

for x=1 to 8            'send 8 bits through a single I/O pin
toggle 0                ‘change the bit - this is the data bit
high 1                  'clock this bit into the '164
low 1
next                    'clock in a total of 8 bits

debug "I just loaded 01010101 into the shift register"
debug cr
pause 2000
high 3                  ‘latches 8 bits into the first '374

debug "I just latched the data into to the first '374 outputs"
debug cr
pause 2000

low 3                   'disable the first '374 latch again

pause 4000

'This is for the second '374

low 2                   'clear the '164 to 0000
```

```
high 2                  'enables the '164 to receive a new value
high 4                  'latch 0000 into the '374
low 4                   'disable the second '374

debug "I just cleared the register and the latch to 0000"
debug cr
pause 2000

for x=1 to 8            'send 8 bits through a single I/O pin
toggle 0                'change the bit - this is the data bit
high 1                  'clock this bit into the '164
low 1
next                    'clock in a total of 8 bits

debug "I just loaded 01010101 into the shift register"
debug cr
pause 2000
high 4                  'latches 8 bits into the second '374

debug "I just latched the data into the second '374 outputs"
debug cr
pause 2000

low 4                   'disable the second '374
debug cls
pause 4000
goto here
```

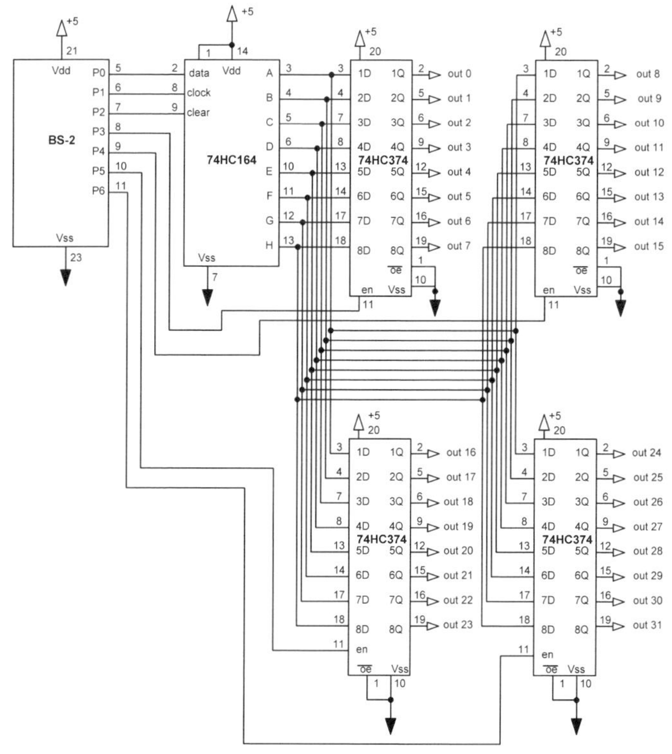


**Figure 5.10**

***32 individually addressable output lines using only 7 I/O lines***

## Code 5.10

```
x var byte
low 0                  'set data line to 0
low 3                  'disable first '374
low 4                  'disable second '374
low 5                  'disable third '374
low 6                  'disable fourth '374
```

```
here:

'This is for the first '374
low 2                   'clear the '164 to 0000
high 2                  'enables the '164 to receive a new value
high 3                  'latch 0000 into the '374
low 3                   'disable the latch again

debug "I just cleared the register and the '374 to 0000"
debug cr
pause 2000

for x=1 to 8            'send 8 bits through a single I/O pin
toggle 0                'change the bit - this is the data bit
high 1                  'clock this bit into the '164
low 1
next                    'clock in a total of 8 bits

debug "I just loaded 01010101 into the shift register"
debug cr
pause 2000
high 3                  'latch 8 bits from the '164 into the '374

debug "I just latched the data through to the first '374"
debug cr
pause 2000

low 3                   'disable the first '374 latch again

pause 2000


'This is for the second '374
low 2                   'clear the '164 to 0000
high 2                  'enables the '164 to receive a new value
high 4                  'latch 0000 into the second '374
low 4                   'disable the latch again

debug "I just cleared the register and the '374 to 0000"
```

```
debug cr
pause 2000

for x=1 to 8            'send 8 bits through a single I/O pin
toggle 0                'change the bit - this is the data bit
high 1                  'clock this bit into the '164
low 1
next                    'clock in a total of 8 bits

debug "I just loaded 01010101 into the shift register"
debug cr
pause 2000
high 4                  'latch 8 bits from the '164 into the '374

debug "I just latched the data through to the second '374"
debug cr
pause 2000

low 4                   'disable the first '374 latch again

pause 2000

'do the same for the third and fourth '374 and strobe data
'thru using P5 and P6 I/O lines

goto here
```

# Analog to Digital Conversion

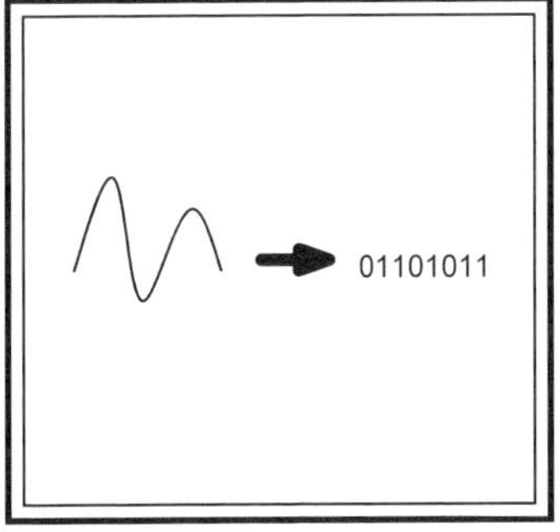


Simply stated, analog means that there are many conditions or levels that a device can assume. For example, you could think of a door as either being "open" or "closed." In a perfect world this would be a digital (or a binary) condition.

In reality, however, there are many different conditions (or "states") in which the door could be. Yes, it could be "open" or "closed," but it could also be *partially* open.

In fact, there are an infinite number of positions in which the door could be. It could be "open" 1.12," or open 1.00014," etc. Although we would like to think of the door as "open or closed," in the "real world" we can't, because it could be open in varying degrees.

This presents some interesting challenges to a microcontroller (or for that matter, any digital device) because they only recognize two distinct "states" - 0 or 1.

Looking at the "real world" door example, we could assume that the closed state of the door is considered a binary "0." Likewise, the full open state could be considered a binary "1."

If we connect some input sensors (such as contact switches) to the doorframe and then connect these switches to a micon, it would be able to detect when the door was fully open or fully shut.

However, in the time between when the door leaves the fully shut condition and travels to the fully open condition, the microcontroller is unable to detect or "follow" the door *as it is in the process of opening.* With the exception of the two end points (during the swing of the door) there is no data available to the micon – it's blind.

What we need to do is create an electrical interface that allows the microcontroller to follow the progress of the door as it is opening. Then, not only can the micon verify that the door was successful in opening all the way, but it would also have the ability to monitor and control how far the door should be opened. For example, your system could open the door 30% for "ventilation."

When we create the circuitry that enables a micon to see the current state of an analog condition, we're doing *Analog to Digital Conversion*.

In Figure 5.11, we've attached a potentiometer to the +analog voltage input of an A/D converter. Any analog input signal between 0 and +5 volts can be applied here.

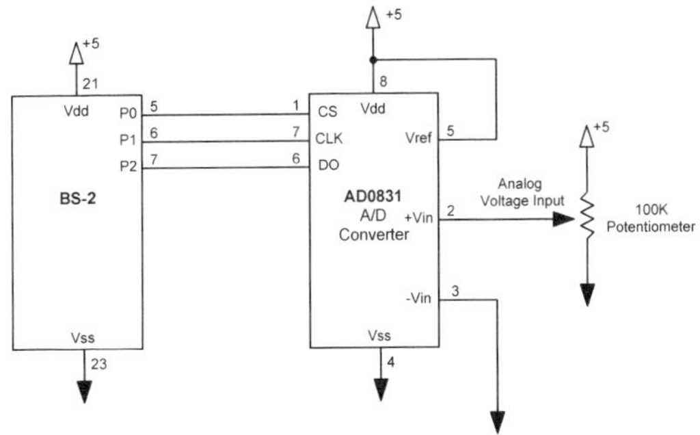


**Figure 5.11**
***A simple analog to digital converter***

**Code 5.11**

```
x var byte
y var byte
here:
low 0              'select the converter
pulsout 1,1        'send the first "setup clock pulse"
x = 0              'set x to zero
for y  = 1 to 8    'loop 8 times to get the 8 data bits
pulsout 1,1        'send a clock pulse
x = x * 2          'shift bits once to the left
x = x + in2        'add x to the incoming bit
next               'do it 8 times
high 0             'de-select the ad0831
debug ? x          'print the result
goto here
```

The AD0831 is an 8 bit "successive approximation" analog to digital converter. It is connected to a microcontroller using only three control lines.

Chip select is an active low input. To begin the conversion process, this line is taken low.

"DO" is the data line. Each time "clock" (pin 7) is toggled, DO provides the next bit in the serial data. "Clock" must be toggled once to initiate the conversion process, then each subsequent pulse produces the next bit in the converted value.

# A/D Span Circuitry

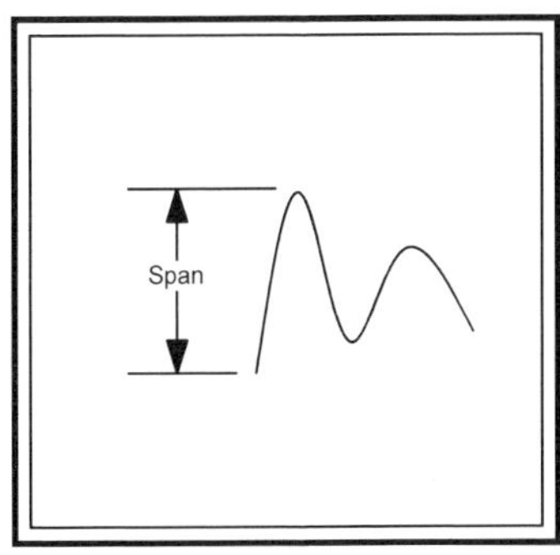


*Span* is the range of voltage over which an A/D converter will measure.

For example, in Figure 5.11 the span was 5.0 volts - the converter returned a value of "0" when the input voltage was at ground potential, and it gave a result of 255 when the detected voltage was at about 5 volts.

Let's suppose for example, that the voltage you wish to measure is between 0 and 2.5 volts. In order to get maximum resolution from the converter, we need an output of "0" when the voltage is at 0, and 255 when it reaches 2.5 volts.

By using the circuit as shown in Figure 5.12, you can adjust the *span* (in this example, 2.5 volts - the difference between 0 and 2.5 volts),

By narrowing the scope over which the converter measures, we reserve the same number of "steps" (256 in an 8 bit converter) to be spread over a smaller voltage range - effectively increasing our resolution.

Because devices like the AD0831 are designed to be "adjustable," we can alter the span over which the converter operates by simply adding a couple of resistors.

The AD0831 has a pin called "$V_{ref}$" that can be biased in such a manner as to adjust the span of the device. $V_{ref}$ is located on pin 5. As you can see in Figure 5.11, we connected $V_{ref}$ to +5 volts. Since this biases pin 5 at 5 volts, the span was set at 5 volts.

If we can apply (or bias) a voltage other than 5 volts to this pin, then that bias voltage will determine the span or "range" of the converter.

To accomplish this we're going to use a simple voltage divider. Refer to Figure 5.12.

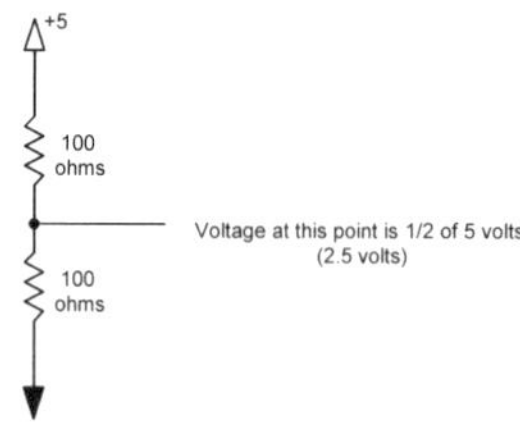


**Figure 5.12**
***Using two resistors to create a voltage divider***

**Code 5.12**

"no code"

Notice that if we attach two resistors (in series) between Vss and Vdd (5 volts) that there will be a current flow through both of these resistors. If you measure the value at Vdd (in reference to ground), you should get about 5 volts.

Now, measure the voltage at the junction of the two resistors. Since the values of the resistors are equal, the voltage (at this junction) will be approximately ½ that of Vdd.

By changing the value of the "top" resistor to 200 ohms, the voltage (at the junction) will be about 1/3 of Vdd. As you continue to increase the value of the "top" resistor, you can see that the voltage at the junction of the two resistors will get lower and lower. This is the basic operation of a voltage divider. We're "picking off" a voltage that has been *divided* to a certain value.

Figure 5.13 shows a bias (on Vref) of 1/2 Vdd, or approximately 2.5 volts. The digital output will now go from 0 to 255, over a limited span of only 2.5 volts. You can change the value of the bias voltage to suit whatever span your project requires.

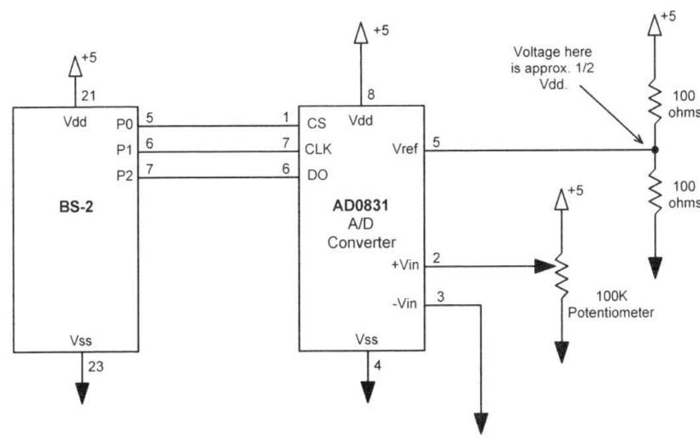


**Figure 5.13**
***Span adjustment of 2.5 volts on the AD0831***

**Code 5.13**

"same as code 5.11"

# A/D Zero Offset Circuitry

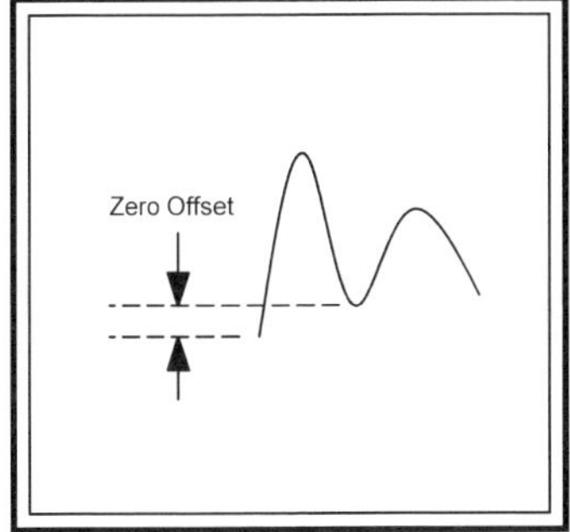


Figure 5.13 altered the span of a converter so that it can be scaled to accommodate different *ranges* of input voltages. However, the span to which the converter was referenced was fixed at 0 volts ("ground").

What happens, however, if you have an analog voltage that you wish to measure that is between say, 2.5 and 5.0 volts? In this situation, we need to change the "zero" point (the voltage level at which the converter gives us a digital value of "00000000" as an output).

If the analog input voltage is anywhere between 0 and 2.5 volts, the converter gives us a "zero" reading. As the analog voltage continues to increase from 2.5 up to 5.0 volts, the AD0831 makes the conversion (based on it's 256 steps of resolution), and produces the appropriate digital value.

This is what is called "zero offset" - the ability to move the zero point of the converter to whichever level is most appropriate for your application.

In the prior A/D circuits, we've had -Vin (pin 3 on the AD0831) tied to Vss. This is the minus side of the analog voltage input pair. Our positive voltage input has been on pin 2 (+Vin). By biasing -Vin (in a similar way as we did for Vref), we can instruct the AD0831 to use whatever voltage (created in our new voltage divider) that is present on -Vin, as the point at which the converter will output a "0" value.

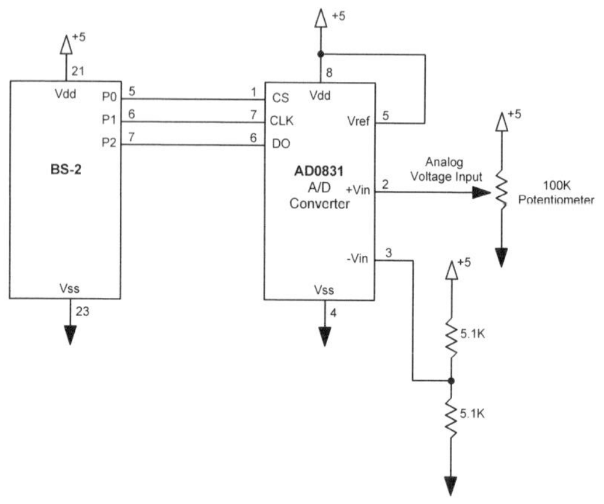


**Figure 5.14**
***Zero offset adjustment of 2.5 volts***

**Code 5.14**

"same as code 5.11"

Note that the resistors connected to -Vin are equal in value (5.1k). Equal resistances in the divider network yield a bias voltage of Vdd divided by two; In this case, 2.5 volts.

Run the program now. If you have a VOM, connect it to +Vin and Vss. Notice that as you increase the voltage (using the pot) that your output on the PC doesn't change from "0" until you go beyond about 2.5 volts. Also notice that the most you can get out of the AD0831 is about "128" (or so), as displayed in the PC's debug window. Why is this?

Remember that the voltage present on Vref determines the "span" of the converter. Since Vref is connected to Vdd, the span is set to 5 volts.

Because we're starting at 2.5 volts (by moving the "zero offset") the converter will output the 256 values over a span of 2.5 to 7.5 volts (2.5 volt zero-offset plus a 5-volt span).

However, since the total voltage available in our circuit is only 5 volts, we will only "see" a conversion over ½ of the available span.

Remove the wire that connects Vref (pin 5) to Vdd, and replace it with two 100 ohm resistors in a voltage dividing network, as shown in Figure 5.15.

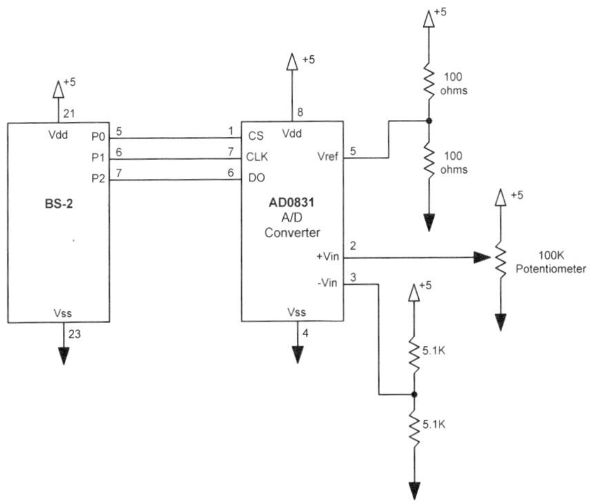


**Figure 5.15**

***2.5 volt zero offset, combined with a 2.5 volt span***

**Code 5.15**

"same as code 5.11"

Rotate the pot. The converter will now be able to measure any voltage between 2.5 and 5.0 volts with 256 “steps” of resolution.

# Digital to Analog Conversion

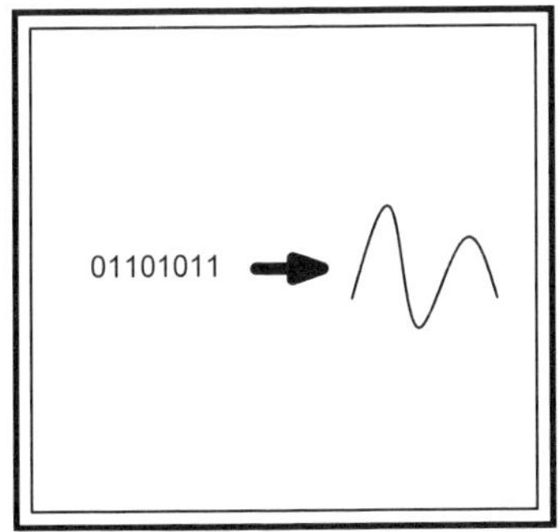


By now, we're well aware that microcontrollers are "digital devices."

The world, however, is rarely "binary" in nature. There are many instances where a microcontroller must know what a particular analog voltage is on a given device or sensor. By using an A/D chip we were able to convert an analog voltage to a digital representation that a micon could understand.

What about going the other way? Although it may not be as common, there are many situations in which you might want to have a digital system generate an analog voltage. Examples might be to control the intensity of a lamp or the speed of a motor.

A Digital to Analog Converter (DAC) enables a *digital* micon to generate an *analog* voltage.

Figure 5.16 is one of the simplest (and least expensive) methods to create a D/A converter. This circuit is called a "resistive ladder D/A converter." The name comes from the fact that the resistor network (as drawn in the schematic) kind of looks like a "ladder."

This circuit is an 8 bit D/A converter, utilizing I/O pins P0-P7.

If you don't want to use eight bits to create this circuit, you can blend Figure 5.16 with Figure 5.8, which uses a shift register to reduce the number to only five I/O lines.

Aside from doing the opposite of what the AD0831 A/D converter did, this circuit also has other differences.

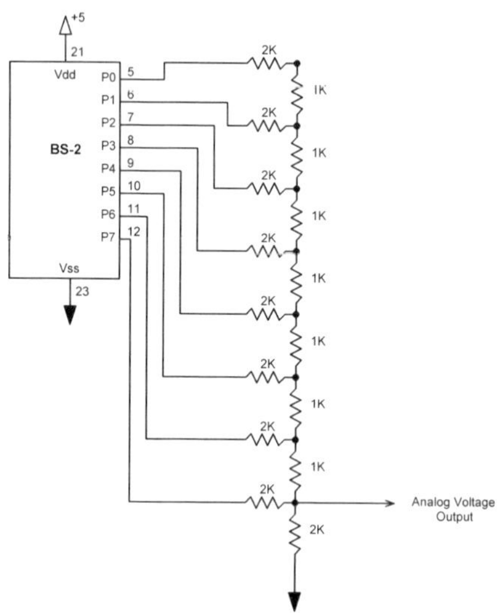


**Figure 5.16**
***A simple resistive ladder D/A converter***

**Code 5.16**

```
x var byte
dirl=255
here:
for x= 1 to 255
outl=x
pause 50
next
goto here
```

The first is the fact that we're using very inexpensive components to accomplish the conversion. Resistors (the type we're using) are cheap - less than a penny each.

Secondly, the number of bits used in the conversion (in this case 8) are all presented to the "converter circuit" simultaneously. That is to say that the binary data to be converted to an analog voltage arrives at the "converter" (portion of the circuit) as *parallel* data.

The AD0831 is a serial device. Recall that we had to "clock" the digital data out of the converter in a sequential manner. The advantage of this is that we only used three I/O lines to bring the data into the Stamp. Serial, however, is considerably slower than an equivalent parallel device because of its sequential nature.

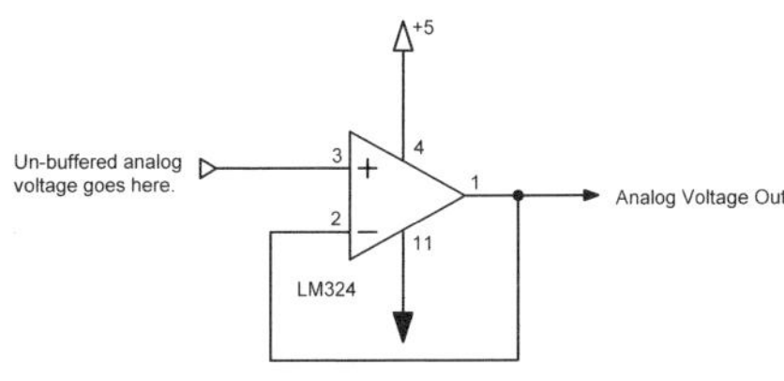


**Figure 5.17**
***A voltage following buffer circuit***

**Code 5.17**
"no code"

Connect a VOM to the "Analog Voltage Out", and run Code 5.16. The voltage output will ramp up and then recycle at the rate determined by the "Pause" command. If you're using a digital VOM it might be easier to see the voltage change if you increase the length of the "Pause".

The circuit in Figure 5.16 is quite sensitive to loading. That is, if you try to use the D/A converter to drive another circuit, the analog voltage may drop (depending on the load of the circuit.)

To prevent this, you can add an analog buffer circuit as shown in Figures 5.17 or 5.18.

Figure 5.17 has the advantage of operating from the same single-ended power supply, but suffers from not being able to provide a voltage output any higher than about 3.5 volts (due to limitations within the LM324).

Figure 5.18 will provide a much greater voltage output (to Vdd), but requires a double-ended power supply.

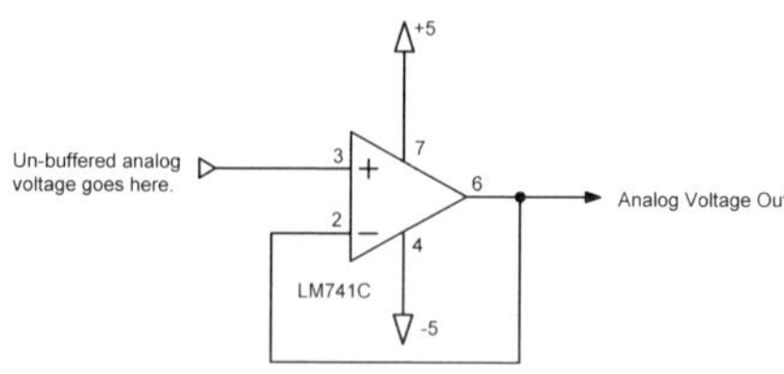


**Figure 5.18**
***Analog voltage-following buffer***

**Code 5.18:**
"no code"

The accuracy of a typical "home-brew" resistive ladder DAC is primarily dependent upon the tolerances of the resistors used.

## Appendix

# Parts Sources

# Parts Sources

Parallax, Inc. 1-888-512-1024
www.parallaxinc.com
www.stampsinclass.com

Simply, *the* source for Stamp related stuff. Many of the parts used in this book are available here as well.

Mouser Electronics 1-800-346-6873
www.mouser.com

Comprehensive selection of electronic components. Free catalog on request. Entire catalog available on website. No minimum order requirements and they ship same day with no handling charges. No voice-mail – they actually answer the phone. Complete tech and quote department – provides cross-referencing services and specification sheets.

Allied Electronics 1-800-433-5700
www.alliedelec.com

Another source for electronic components. Free catalog and CD-ROM on request.

Jameco Electronics    1-800-831-4242
www.jameco.com

Electronic components as well as computer parts. Free catalog.

Digi-Key    1-800-344-4539
www.digikey.com

Another basic electronic component parts source. Free catalog.

Radio Shack    1-800-THE-SHACK
www.radioshack.com

They're everywhere. Basic selection of component parts at their retail stores, but through "Radio Shack Unlimited", they can direct ship much more to your door.

Edmund Scientific    1-856-573-6250
www.edmundscientific.com

Nifty devices for optical applications (lenses, cameras, lasers, etc.)

WM Berg    1-800-232-BERG
www.wmberg.com

*All kinds of mechanical drive components, etc.*

McMaster-Carr 1-630-833-0300
www.mcmaster.com

*An almost unbelievable assortment of mechanical devices – motors, solenoids, etc. Catalog has an extensive index. Call the above number for their local distribution center and phone number.*

## Other information sources:

Check out my web-site at **www.miconcookbook.com** for information and updates to this book.

*Programming and Customizing the Basic Stamp Microcontroller*, by Scott Edwards
ISBN 0-07-913684-2

*The National Electrical Code*
ISBN 0-442-02223-9

*Wiring Simplified* by Richter and Schwan
ISBN 0-9603294-4-7

Parallax's website at www.stampsinclass.com has a large amount of free (downloadable) learning materials such as :

*What's a Microcontroller?*
*Robotics*
*Environmental Projects*
*Analog and Digital*

These materials provide a great foundation for exploring the world of the microcontrollers, especially if you're just starting out.

This is also where you can download a complete copy of the BASIC Stamp Manual.

Be sure to join the discussion forums here too. And if you're new to microcontrollers (especially the Stamp), don't be afraid to post questions. There are many kind people here who love to enlighten those of us who are interested in learning more about this technology.

# Index

# Index

## M



## N



## O



## P

## V



## W



## Z